Elements to Elephants

Life becoming, diverse

Dr Matthew Sawyer

ISBN-10: 1536947776
ISBN-13: 978-1536947779

To Amelia

for asking the question

Contents

Part 1 - Cells

1	Introduction to cells	page 1
2	Storing and using information	page 21
3	Genes working right and wrong	page 49
4	Mitochondria	page 65

Part 2 - Evolution

5	Evolution	page 85
6	HIV - Evolution and Effects	page 115
7	Bacteria and Virulence	page 135
8	Bacteria and Humans	page 153
9	Archaea	page 183

Part 3 - Biodiversity and ecosystems

10	Biodiversity	page 199
11	The soil ecosystem	page 227
12	The marine ecosystem	page 253
13	The woodland ecosystem	page 277
14	The future of genes and diversity	page 295

Glossary page 314

Index page 323

Acknowledgments

Firstly to Becky, who believed from the very first word. To Nick for helping to edit the words into a more coherent order and to Emily for the illustrations. To Ryan for the loan of the books and to Alex, Sam and Vicky for proofreading and offering feedback. Last, but by no means least, a big thank you to three inspirational and enthusiastic lecturers at Teesside University - Helen, Caroline and David - without whom this could never have happened.

Nick Hodgson can be reached via www.root-and-branch-editing.com

About the author

Dr Matthew Sawyer is a working medical doctor with a keen interest in the natural world and environmental issues. He is currently using his free time to pursue a degree in Environmental Science. Matthew finds microorganisms particularly exciting much to the consternation of his friends and colleagues. He has one chicken called Darwin.

Introduction

This book is aimed at those who have a curiosity about the origins of life, evolution and interactions within the environment. Suitable for those from GCSE to A-level, degree level and beyond, the book covers three main areas.

Firstly, the initial building blocks of cells and their subsequent development are explored. The three main molecules of life - DNA, RNA and proteins - are examined along with the importance of energy and its storage. How cells are able to turn genes on and off is shown in detail.

Secondly, cells evolved and developed into more complex organisms. Darwin with his finches developed several basic principles which show how organisms have changed in response to different environments. Using HIV and antibiotic resistance as examples, evolution of organisms in our lifetime is demonstrated. The interaction between bacteria with mankind is reviewed including several diseases. The development of antibiotics is reviewed as is the mechanism of antibiotic resistance. The least well-known group of organisms - the archaea - are explored in a chapter of their own.

Finally, an introduction to ecological theory covering soil, oceans and woodlands. This includes a deeper understanding of the different factors influencing the development and survival of different organisms. Simple biodiversity calculations are worked through. The importance of maintaining a wide variety of organisms and their genes is emphasised.

The concluding chapter examines some of the current and future developments within gene technology and maintaining biodiversity. This includes how genes are currently used and how they could be manipulated in the future.

The main passages in this book are arranged so that they can (hopefully) be understood by all. More advanced readers should be able to understand the 'more details' sections, but feel free to skip ahead.

element, *n.*

A component part of a complex whole.
One of the simple substances of which all material bodies are compounded.
"element, n." *OED Online*. Oxford University Press, June 2016.

elephant, *n.*

A huge quadruped of the Pachydermate order, having long curving ivory tusks and a prehensile trunk or proboscis...the largest living land animal.

'to see the elephant' (U.S. slang): to see life, the world; to gain knowledge by experience.

"elephant, n." *OED Online*. Oxford University Press, June 2016.

Chapter 1 - An introduction to cells

A cell is regarded as the true biological atom.

- George Henry Lewes

What are cells?

Cells make up all living organisms. Everything we consider to be alive has one or more cells; they are termed either single-celled or multi-celled. A cell has different parts inside it to do different jobs.

Animal and plant cells are a bit like a big cardboard box with other smaller boxes inside. The big box is the cell itself, while the cardboard exterior is the cell wall or membrane. On the outside of the box are labels to tell you what is inside. For example, if you are moving house, they may be labelled 'kitchen' or 'bedroom'. If the box is to be posted, the label may say 'fragile' or 'this way up' along with the address label. Inside the main box are smaller boxes, as in Fig. 1. In our case, one smaller box contains a recipe book, another contains a hand-wound torch, another some beads, and another a necklace or bracelet maker. If this box were representing a plant cell, there would also be a box with a solar panel inside. Each of these internal boxes contains specialist kit to do a specific job role for the cell and are called organelles.

The main box represents the outer cell membrane. The labels represent the proteins by which the cell is recognised. The recipe book stores information and represents DNA. The hand-wound torch generates energy and represents the mitochondria - often called the 'powerhouse' of the cell. The beads represent amino acids, which are the building blocks of proteins. The bracelet maker pieces together the amino acids and is called a ribosome. The workings of each part will be explained in more detail later in the book.

Fig. 1 Boxes inside a box

Picturing a cell as a box is a useful mental image but, as we will see, a cell is a little more complex....

The outside of a cell

Let's start with the cardboard box. Corrugated cardboard is made up of two layers of cardboard with a thin layer of paper sandwiched in between, as we can see in Fig. 2. This layering gives the box strength and rigidity and makes it useful for carrying many things inside it.

Fig. 2 Two-layered cardboard

The walls of a cell are not made of cardboard but from molecules called phospholipids ('phospho' = phosphate containing; 'lipid' = fatty or oily). A phospholipid has a phosphate 'head' and a fatty 'tail' and is shown in Fig. 3. The head is hydrophilic ('hydro' = water; 'philic' = having an affinity for or loving) and the tail is hydrophobic ('phobic' = hating).

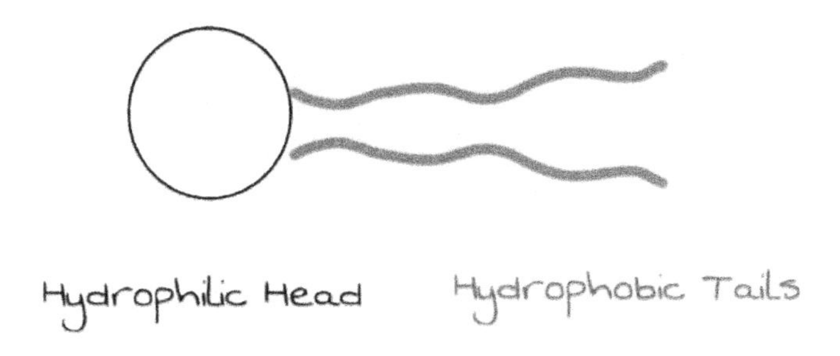

Fig. 3 Diagram of a phospholipid molecule

If cooking oil and water are put in a jar and shaken vigorously (put the lid on tight!), the two liquids are destined to remain apart forever. The oil starts to clump together in little balls or spheres that gradually join and coalesce with other oily balls until it returns to one single layer of oil. In a phospholipid molecule, the tail is a 'fatty acid' and fat, as we have seen, does not get on with water. As a result, the tails are forced to clump together in the presence of water, and fatty acids cannot be broken down by water.

More details

The fatty acid chains are made of strings of carbon and hydrogen atoms. Atoms find peace when they are resting or are at their lowest energy levels. For a carbon atom, this lower-energy state occurs when it is sharing four electrons in four bonds. In a fatty acid, two of the bonds are to two hydrogen atoms; the other two bonds are to two different carbon atoms. Overall, this forms a carbon backbone with hydrogen atoms protruding from each carbon atom, as seen in Fig. 4.

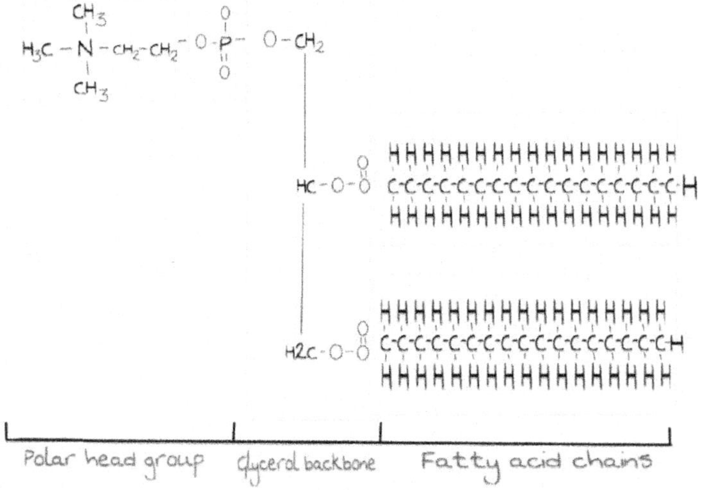

Fig. 4 The chemical structure of a fatty acid

If each carbon atom were joined to two carbon atoms and two hydrogen atoms, a long linear molecule would be produced, forming the tail of the phospholipid. Sometimes, however, rather than a chain of carbons with a single bond (-C-C-C-C-), a double bond occurs (-C-C=C-C-). In this sequence, the first and fourth carbon atoms have two hydrogen atoms attached – meaning there are four bonds. As the second and third carbon atom already have three bonds, they can have only one more hydrogen atom attached to each of them. A 'double bond' creates an angle and changes the shape of the fatty acid from a straight line to one with a kink in it. This explains the difference in shape of a molecule between a 'saturated' fatty acid, which is saturated with hydrogen atoms, and an 'unsaturated' fatty acid with carbon double bonds, as seen in Fig. 5.

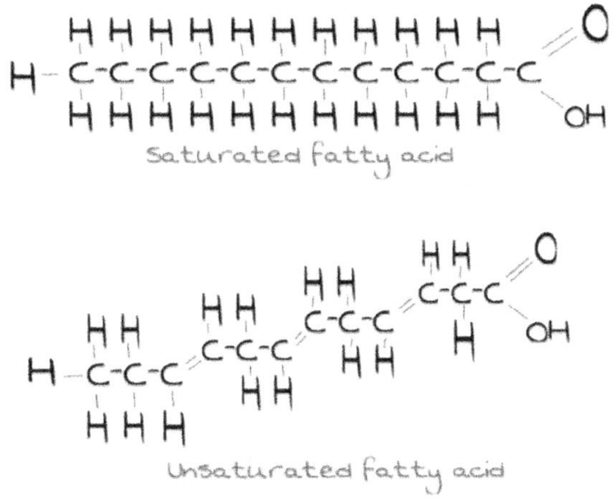

Fig. 5 A saturated fatty acid molecule forms a straight line; an unsaturated fatty acid molecule forms a kinked shape

The shape of the fatty acid depends partially on the number of carbon atoms with double bonds but also on the length of the chain.

Building the cell wall

If these phospholipid molecules are dropped into water, just as with oil, they try to keep the water-hating tails away from the water molecules as best they can. They organise themselves into the simplest of shapes, forming a ball with the hydrophobic tails pointing towards the centre. This is called a micelle and is seen in Fig. 6.

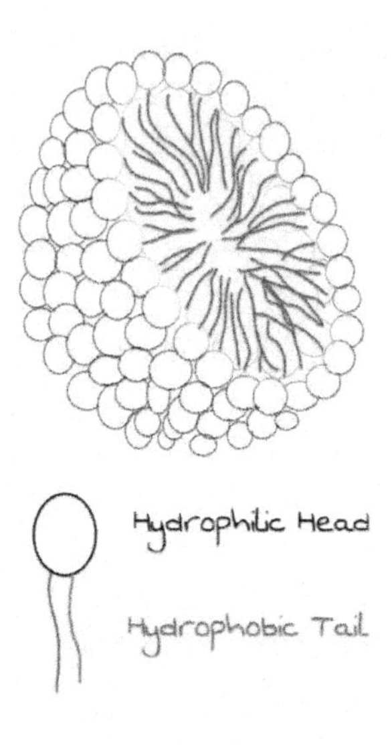

Fig. 6 A micelle formed from simply organised phospholipid molecules

If there are more phospholipids, more complex shapes can be made. Rather than a simple ball shape with all the tails pointing inwards towards the centre, a shape develops that has two layers of molecules with the heads facing both inwards and outwards simultaneously. This is the start of the phospholipid bilayer ('bi' = two) of the cell membrane seen in Fig. 7. These molecules move over and past each other as if made of a fluid themselves. In a cell membrane, the chemical properties of the phospholipid bilayer allow some substances, such as gases like oxygen or carbon dioxide, to pass through by diffusion. A simple sphere with the tails together and a layer of 'heads' facing the centre is called a liposome.

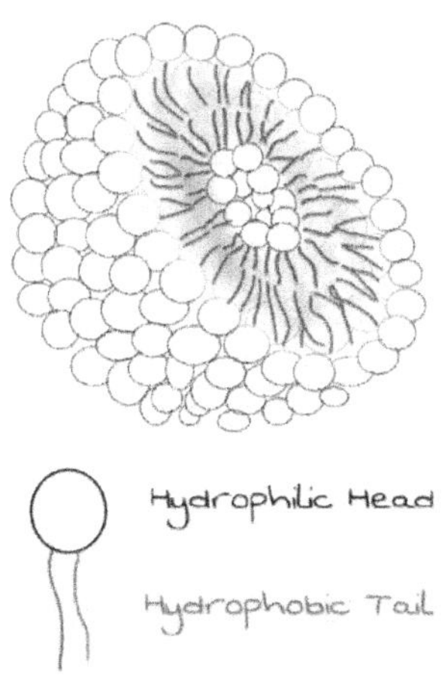

Fig. 7 Organised phospholipid bilayer as a liposome

As the sphere becomes bigger and bigger, the area in the centre enlarges. Eventually, this becomes a fluid-filled space surrounded by two layers of phospholipid molecules arranged to keep water away from the fatty tails. If other molecules - such as protein or sugar molecules - are trapped in this space, the very early beginnings of a cell appear.

Is size important in a cell?

As this sphere grows in three dimensions, the centre gradually gets further and further away from the outer walls. Imagine blowing a soap bubble using a straw and soapy water. To start with, the soap coats the end of the straw in a thin layer. There is no bubble but, as air is blown down the straw, a bubble appears and gradually enlarges. The larger the bubble gets, the further the outer walls are from the centre.

Why is this important? In a cell, the waste products of food such as carbon dioxide move by diffusion. This is the movement of a substance down a concentration gradient from a place with a large amount to a place with a smaller amount. If food and nutrients are needed in the centre of the cell or when waste gases are produced, the larger the cell the longer it takes for them to enter or exit the cell. This correlates to the surface area to volume ratio. When an object is small, it has a large surface area to volume ratio. As it grows, this ratio falls, and this affects the diffusion rate. As a result, there is a finite size a single cell can grow to until it can no longer successfully move nutrients or waste in or out.

Think of a struggling swimmer in a pool. The smaller the pool, the easier it is for the swimmer to reach the side or be rescued. However, if the body of water were too big - a large lake, for example - it would be much more difficult to save the beleaguered swimmer or for them to reach dry land.

More details

Visualise a cube 1cm x1cm, as presented in Fig. 8. To calculate the area of each side, the equation is length x breadth = area. This works out as 1cm x 1cm = $1cm^2$ and there are six sides of $1cm^2$ each, or a total surface area of $6cm^2$. The volume of the cube is length x breadth x height or 1cm x 1cm x 1cm, which is $1cm^3$. Therefore, the ratio of surface area to volume is $6cm^2$:$1cm^3$, or 6:1.

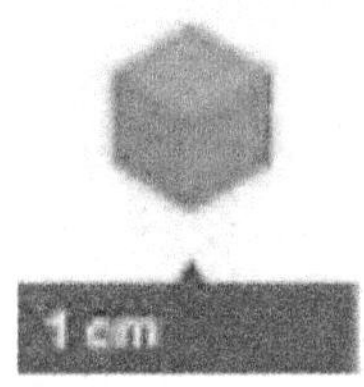

Fig. 8 A one-centimetre cube

If the lengths of each side were doubled to 2cm, as in Fig. 9, using the same equations as before, the area of each side is 2 x 2cm = $4cm^2$. In total, the surface area of the six sides is 6 x $4cm^2$, or $24cm^2$. The volume becomes 2 x 2 x 2cm = $8cm^3$. This ratio is 24:8 and equates to 3:1.

Fig. 9 A two-centimetre cube

If the sides of the cube were lengthened further to 3cm each, the surface area would become 3 x 3cm or $9cm^2$, resulting in a total of $54cm^2$ for the whole cube. The volume would be 3 x 3 x 3 = $27cm^3$; the ratio is now 54:27 or 2:1.

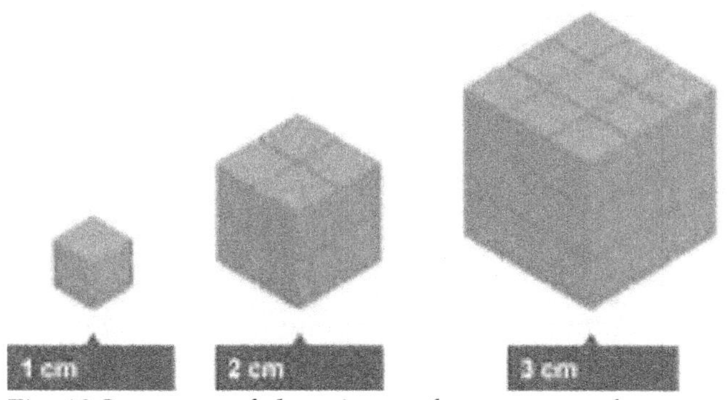

Fig. 10 Summary of changing surface area to volume ratios

We can see in Fig. 10 that the larger the cube, the smaller the ratio of surface area to volume.

Even more details

Physics dictates that (generally) the shape with the smallest energy requirements is a sphere; therefore, cells tend to be spherical.

To calculate the surface area to volume ratios of a sphere the following equations are used.

$$\text{Volume of a sphere } V = \frac{4}{3}\pi r^3$$

$$\text{Surface area of a sphere } S = 4\pi r^2$$

Table 1 demonstrates the falling ratio as a cell's radius length increases.

Table 1. Falling ratio with increasing size

Radius	Volume (V)	Surface area (SA)	Ratio (SA/V)
1	4.19	12.57	3:1
2	33.51	50.27	1.5:1
3	113.1	113.1	1:1

Just as with the cube, there is a fall in the ratio with increasing size.

When a cell reaches a certain critical size, the size constrains it and it can no longer function efficiently. Gases such as carbon dioxide and oxygen cannot diffuse into or out of the cell sufficiently. So what next? One option is for the cell to divide itself in two. It could create two identical daughter cells and start the whole process of enlarging, reaching a critical size, and replicating forever. Another option would be to increase the surface area within the cell to allow gases and nutrients in and out more effectively. Another option would be to organise different parts of the cell to carry out different functions in specialist centres known as organelles. These could be cordoned off from the rest of the cell. The nutrients and waste could then be ferried only to the parts where they are needed.

Increasing the surface area of a cell membrane

The surface area of a cell membrane can be increased in the same way that a radiator at home works. To allow more heat to escape, a radiator has extra folds or metal 'fins' added. This increases the surface area and allows more heat to be transferred from the water inside the radiator to the surrounding room. A cell can increase the surface area available by infolding the outer cell membrane.

Of course, the cell is a 3D structure rather than a 2D picture. In a cell, some infolding of the membrane occurred. This may have been by a 'bubble' or liposome of phospholipid joining the cell, but then NOT collapsing back onto itself as it joined. It may be that there was a slight difference between the lipids; for example, in the number of different bonds there were in the fatty acid chain.

The essential molecules for life

We have seen how a cell wall can develop from phospholipid molecules, but how do these and other chemicals form? To understand this, we need to go back in time.

The Earth is old. It is not as old as the universe - which has been around for about 15 billion years since the Big

Bang and counting - but the Earth was formed about 4.6 billion years ago. It took about 0.6 billion years for the gases to solidify into rock and for the oceans and land masses to form. Surprisingly, it was only about 0.5 billion years after that when the first living cells appeared.

Experiments have shown that atoms join together and bond to form new and bigger molecules and chemicals. Many of these meetings are by chance and occur randomly. How can we know that one atom could bump into another in the right order to create a meaningful molecule? It's down to the large numbers involved, and we have to think big - very big. When we think of generating a random number we could roll a couple of dice or use a random number generator, but the numbers would be too small. The numbers involved in atoms potentially colliding and joining together are breathtakingly large. For example, in just over 2 teaspoons or 12g of carbon (the atomic number of carbon is 12) there are 600,000,000,000,000,000,000,000 atoms. That's 6 with 23 zeros afterwards. This is known as *Avogadro's constant.* It is more commonly written in scientific notation as 6.02×10^{23} and is known as one mole. This is a huge number, but our numbers get even bigger.

The weight of the Earth is calculated to be about 5×10^{24}kg. Two-thirds of this is either iron or oxygen. Using this as a rough guide, we know there are 6×10^{23} atoms in 8g of oxygen (as its atomic number is 8) and in 26g of iron (atomic number 26). Overall, this would give an average of 6×10^{23} atoms in every 17g of the planet.

If there is one mole of atoms in 17 grams, then in 1,000 grams or 1kg there are 58.8 moles (1,000/17 = 58.8). If there are 58.8 moles per kg, multiplying it by the weight of the Earth gives 58.8 x 5 x 10^{24}, which gives 2.94 x 10^{26} moles. To convert this to atoms, multiply this by the number of atoms in a mole: 2.94 x 10^{26} x 6.02 x 10^{23}. This gives (roughly) 1.76 x 10^{50} atoms. That is a 1 with 50 zeros, or 100,000,000,000,000,000,000,000,000,000,000,000,000,000,000,000,000,000 atoms. And that's just on Earth. If we consider the length of time the Earth has been forming - 4.6 billion years - we can see that the likelihood of atoms bumping into each other is extremely high.

Making essential molecules

There is an awe-inspiring number of atoms that can collide and randomly create new molecules. The phospholipids in the cell wall are made from simple atoms - hydrogen, carbon, oxygen and phosphate - combining together to form longer molecules.

What other molecules are needed to construct a cell? The three molecules considered essential to life are RNA, DNA and proteins (made from amino acids). Other simple molecules are carbohydrates (made from sugars) and fats (made from fatty acids).

Chapter 2 explores the details of DNA and RNA, but both involve a ribose sugar. Ribose sugars are simple sugar structures made of carbon, oxygen and hydrogen; they are the backbone of RNA and DNA molecules. Recently, the interactions between ultraviolet space radiation, ice in comets and simple molecules such as ammonia (NH_3) and methanol (CH_3OH) have been shown to produce various substances in space. Back on Earth, other essential molecules of DNA, such as the nucleotides thymine, uracil and cytosine, were produced in 2015 in a simulated space environment. (See Chapter 2 for more on nucleotides.)

Amino acids are the building blocks of proteins and have been found in carbon-rich meteors. Proteins are strings of amino acids and are used to speed up chemical reactions (as enzymes). Long chains are called polymers ('poly' = many; 'mer' = parts). Proteins can form into different shapes such as fibres (long polymers that form bridges between themselves) or ball-shaped proteins.

More details

Proteins are formed by joining the building blocks into polymers. Proteins can combine themselves to form even more complex structures. Each protein is made of a unique combination of amino acids; they each interact with chemicals and molecules differently. Bridges are formed between the amino acids in different ways; it is these bonds that determine the final shape of the proteins.

Amino acids bond to each other in several ways. The strongest bond is called a disulphide bond or disulphide bridge. This occurs mostly between two sulphur-containing groups and is sometimes written as an S-S bond. This is a little like concrete in that a lot of energy is needed to break the bonds.

A weaker bond is called an ionic bond. It forms between the positive and negative charges on the amino acids when they have swapped an electron. These give a 'full' charge and are more like superglue. These bonds come apart with far less force than is needed to break the disulphide bonds.

The third type of bond, a hydrogen bond, is the weakest. This forms by weak electrostatic charges on hydrogen (positive) and oxygen (negative). Although these bonds are the weakest, the sheer number in a protein means they have a huge influence over the final shape and structure. They are something like sticky tape, as they hold the protein together by the vast amount used.

The simplest carbohydrate is glucose. This simple sugar molecule can be called a saccharide. A single sugar unit is known as a monosaccharide ('mono' = single). More complex sugars can be made by adding together monosaccharides to form either disaccharides (two sugar molecules joined together) or polysaccharides ('poly' = many). These are called carbohydrates. The sugar molecules are bonded together in a condensation reaction, which happens by removing a water molecule between adjoining sugar molecules.

Fats are formed by joining together a fatty acid molecule and a glycerol molecule. Glycerol is a simple sugar alcohol made from carbon, hydrogen and oxygen. Again, this involves the removal of a water molecule by a condensation reaction.

The need for energy

Cells need energy. Producing energy in single-celled organisms can be complex and is done in several different ways. Some bacteria, called photoautotrophs, make energy from light ('photo' = light; 'auto' = self; 'troph' = nourish). Others use glycolysis, a process of energy generation using glucose and oxygen. Yet others use fermentation, a method that generates energy when there is no oxygen. We will see bacteria generating energy from chemicals in deep-sea vents in Chapter 12.

If too little energy is made or not moved to the parts of the cell that need it, then the cell cannot make enough proteins to grow or divide. So, as we saw earlier, size is once again important!

Creating a cell

For a cell to exist, a cell membrane is needed. This is made of phospholipids, as we have seen. Simple molecules have formed spontaneously, as we know through discoveries on meteors and scientific experiments recreating the conditions in space.

The fluid within the cell membrane is called the cytoplasm ('cyto' = cell; 'plasm' = fluid or liquid). The cell membrane is needed to collect or concentrate useful substances on the inside. For example, rather than having the same amount of salt or water as the outside world, more can be retained or more disposed of as needed. In the same way we heat our homes, keeping the heat on the inside to keep us warm, the cell keeps useful molecules inside its boundaries too. Later in this chapter we discover that each of the organelles also has its own cytoplasm-like liquid.

The building blocks needed to build a cell are coming together. Simple molecules such as ammonia and methanol, along with more complex ones such as ribose sugars and amino acids, have been formed in space.

We have seen the simple phospholipid chains forming into micelles and then liposomes and the beginnings of increasing the surface area by creating infolding into this outer membrane.

Classifying cellular life

Cells are not all the same. Often they don't look the same or act the same. Cell types can be categorised in different ways, much as we can classify transport as cars or lorries, boats or aeroplanes, or group them by colour and have all red or blue or white transport in their own groups.

One way to classify living organisms is by how they store their information.

Current thinking splits cellular life into one of three groups or domains: bacteria, archaea and eukaryotes.

Bacteria are single-celled organisms. They do not have a nucleus and are classified as prokaryotes ('pro' = before; 'karyo' = nut or kernel). We will meet bacteria again in Chapter 7. Archaea are a little more complicated; they are mainly single-celled and do not have a nucleus and so are prokaryotes. However, they do not act like bacteria, as we will see in Chapter 9. Creatures we see around us such as animals, plants and fellow humans are multicellular. These are the eukaryotes ('eu' = well; 'karyo' = nut or kernel) and have their information on life - their DNA - stored in a separate part of their cell called a nucleus. Evolution shows that the prokaryotes came first and some evolved into multicellular organisms.

The infolding of the membrane we saw earlier increased the surface area available for energy generation. As a result, the cell that produces more energy for replication is likely to have an evolutionary advantage over its neighbours and fellow competitors for resources. Some of the infolding of the cell membrane developed into a structure known as the endoplasmic reticulum (ER). The ER forms a continuous network of tubing that is connected to the nuclear membrane surrounding the DNA. In Chapter 3 we discover that ribosomes are attached to the ER and are the manufacturers of proteins.

The cell has started to change appearance - from one without internal structures - a prokaryote - into an early eukaryote. The infolding of the membranes increased the surface area, but as it did so the energy constraints became more important. Energy production was not particularly efficient in early prokaryotes. The best way to make energy is through the generation of a molecule called ATP. In eukaryotes this is done by the mitochondria, as we will find out in Chapter 4. In some bacteria, energy production is done across the cell membrane but is less efficient due to the lack of folded membrane.

Out of the millions of prokaryotic cells on the planet, some had the good fortune to develop an infolded membrane. Over time, this folded internal membrane engulfed some structures such as DNA to create a nucleus. This was the development needed to change from a cell with no internal structures (i.e. a prokaryote) into a cell with internal compartments known as organelles. This cell is the eukaryote.

There are many parts to a eukaryotic cell. The basic structure is shown in Fig. 12. The organelles of the eukaryote allow different parts of the cell to become specialised to do just one job, such as keeping the DNA safe or generating energy. The intracellular components of prokaryotes are spread throughout the cytoplasm of the cell and are more freely distributed rather than being organised into separate compartments.

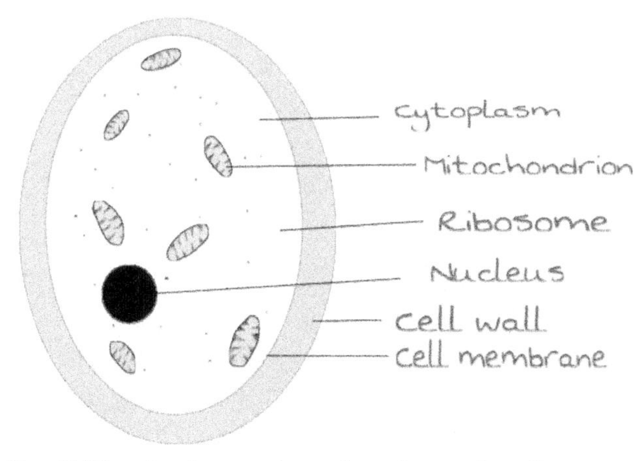

Fig. 12 Very basic structure of a eukaryotic cell

Summary

Life started with atoms forming simple chemicals. These chemicals gradually combined to become more complex. After water-hating fatty acids started forming small droplets, chemicals could become more or less concentrated within the shapes that had been formed. Over time, these became the prokaryote. As life developed more complexity, three groups or domains of cells have been recognised. Those without special storage systems and internal structures are prokaryotes. Those with are eukaryotes. As the cells became bigger, the ratio of the amount of surface cell membrane decreased. Even with the infolding of membranes, more energy was required to run the cell and eventually led to new structures appearing.

Chapter 2 - Storing and using information in a cell

[DNA] ... structure has novel features which are of considerable biological interest.

- Watson and Crick

What is DNA?

For a cell to pass on information to its offspring, it needs to have a way to store the vital knowledge and to copy it. De-oxy-ribo-nucleic acid (DNA) is essentially a way of storing information. It consists of long chains or strands of data that the cell can use like a recipe book when it needs to make things to ensure the smooth functioning of the cell.

Just as a recipe book 'stores' many recipes for starters, main courses, and puddings, biscuits and cakes, DNA does something similar. When you need to make lasagne, for example, you can go to the recipe book, find the ingredients and follow the instructions. DNA does the same thing, only the names of the parts are labelled differently. If the genome is the entire DNA of a cell or the entire book, then, just as books are divided into chapters, the genome is divided into chromosomes. Genes are the sentences within each chapter. Humans have two copies of each book in each cell: one from our mother and one from our father. Sometimes important differences in the information stored in each book are apparent, and one copy can be seen as more important or 'dominant' over the other - as Mendel will explain in Chapter 5. In total, humans have 23 pairs of chromosomes, so a total of 46 in each cell.

Information is stored in a book as letters and words or on a computer as binary code (two bits of information as 0s and 1s). There are 26 letters in the Latin alphabet we use in the West, whereas DNA has far fewer 'letters', using only four to code all the information we need. These four letters are A, C, G and T, representing the four nucleotides of adenine, cytosine, guanine and thymine. Together there are about three billion 'letters' in our human DNA. Enough information for a great many recipes!

What is a nucleotide?

A nucleotide is a ring consisting of five carbon atoms and one oxygen atom, as seen in Fig. 1. As there are five atoms in the ring, it is known as a pentose ring ('pent' = five). The ring itself has four carbon atoms and one oxygen atom, with the fifth carbon atom being a branch from the fourth. To help scientists work out which part of the ring is being talked about, the carbon atoms are numbered as on a clock.

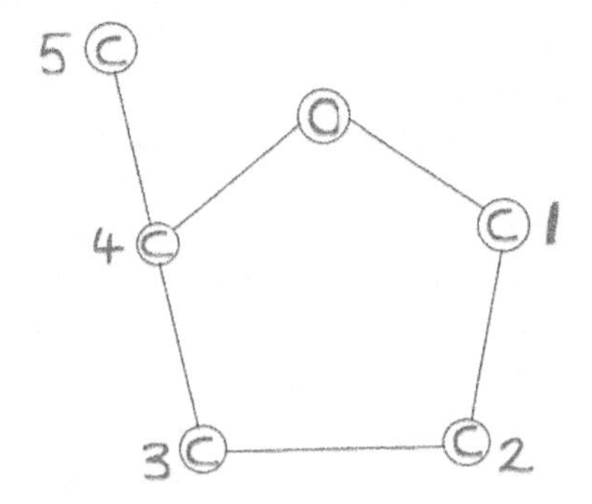

Fig 1 Diagram of a pentose ring (O = oxygen; C = carbon)

To work towards becoming a nucleotide, we need to make the pentose ring into a ribose sugar. We do this by adding more hydrogen and oxygen atoms to the carbon atoms, as in Fig. 2.

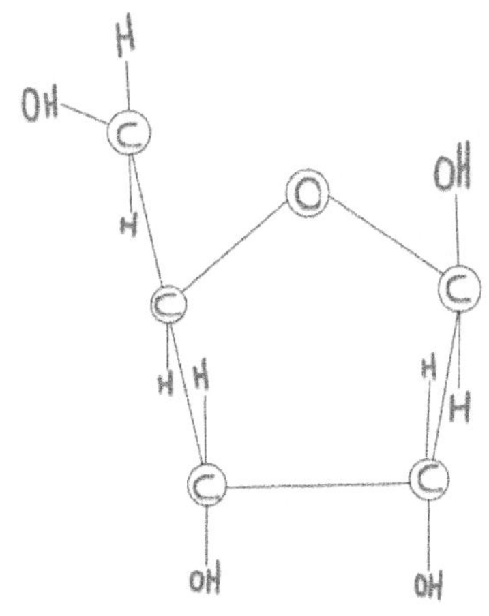

Fig. 2 A ribose sugar molecule

DNA is a chain of ribose sugars that construct the backbone of the molecule. The ribose sugars are joined by phosphate groups, which alternate with them, creating the chain shown in Fig. 3. The beginning of the 'sugar-phosphate' spine starts to take shape.

Fig. 3 Sugar phosphate backbone

The long chain of ribose rings connected together by their phosphate groups is called a ribonucleic acid. Deoxyribonucleic acid (or 'de-oxygenated' ribonucleic acid) is a similar structure, but has lost an extra oxygen atom during the process of the formation of the bonds.

More details

The phosphate group consists of one phosphate atom, three oxygen atoms and one hydrogen atom. The group is attached to the fifth carbon atom of the ribose sugar molecule. It is able to bond onto the third carbon atom of the next ribose sugar by a phosphodiester bond. This bond is formed by a condensation reaction, which is a way of saying that a water molecule is removed. The hydroxyl group (-OH) of the third carbon atom in the ring joins to the -OH of the phosphate group, removing

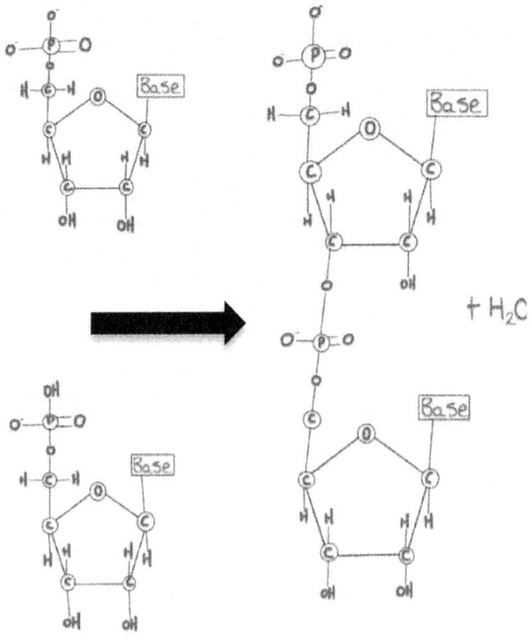

Fig. 4 Formation of phosphodiester bond

H_2O and leaving the oxygen molecule bond to the phosphate atom and the carbon atom of the next ring. This is shown in Fig. 4.

Bonds continue to be made that extend the ribose-phosphate backbone, as shown in Fig. 5.

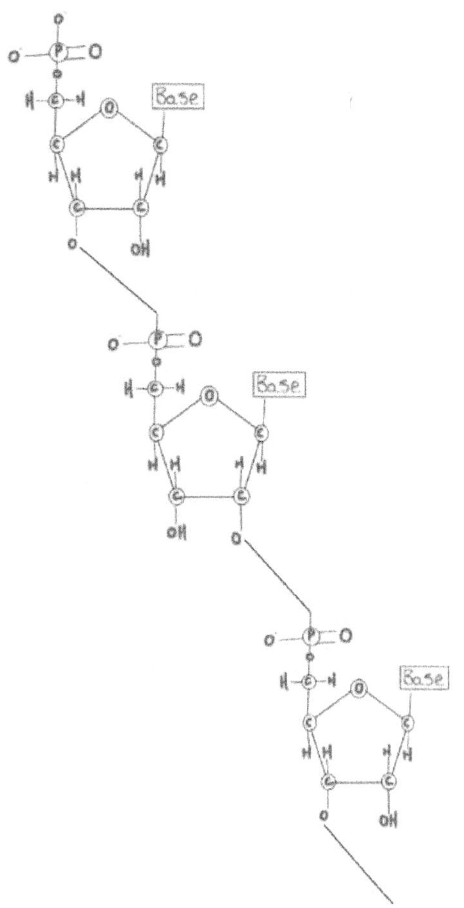

Fig. 5 The bonding of ribose sugars to phosphate groups

Deoxyribonucleic acid has lost an oxygen atom from the second carbon atom and, strictly speaking, is known as 2-deoxyribose.

Making a nucleotide

Now to make a ribose sugar into a nucleotide. We have seen that there are four different nucleotides in DNA. Each of these four is made by adding a 'base' to the pentose ring. Bases are constructed from yet more rings, a bit like those of the ribose sugar, but they also have nitrogen atoms present. Chapter 10 discusses the importance of the nitrogen cycle. For adenine and guanine there are two rings; these are known as purines. For cytosine and thymine there is only one ring; these are called pyrimidines. If we add a base onto the carbon atom in position 1 of the ribose sugar, we have created a nucleotide.

A nucleic acid is a long chain of nucleotides, so RNA and DNA can be called 'biopolymers of nucleotides'. When we have added one of the four bases to each ribose molecule in our polymer it is now called deoxyribonucleic acid - or DNA for short! Fig. 6 shows a diagram of this arrangement.

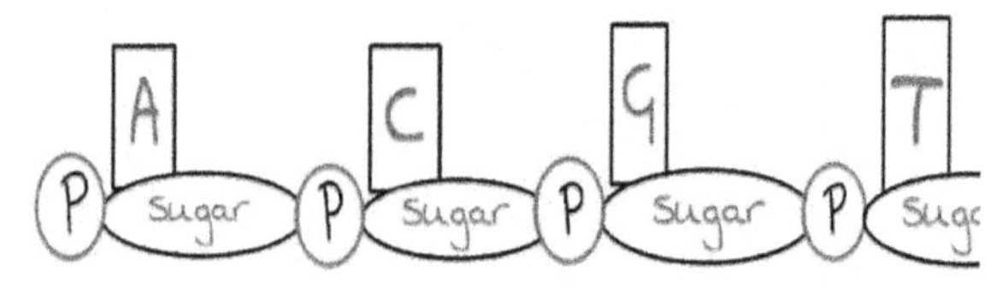

Fig. 6 Sugar-phosphate backbone (deoxyribose) with nucleic acids - DNA!

RNA is slightly different in structure to DNA: it does not use thymine, but a different base called uracil. (See below for more about RNA.)

More details

As Fig. 7 shows the structure of the bases the which is based on rings and how they attach to the first carbon atom of the ribose sugar. Adenine and guanine have two rings: a six-atom hexagonal ring called a hexose ring joined to a pentose ring. Thymine and cytosine have only a single hexose ring.

It shows that the fifth carbon atom (written as 5′) is connected to the phosphate group and then to the third carbon atom (written as 3′) of the next ribose sugar.

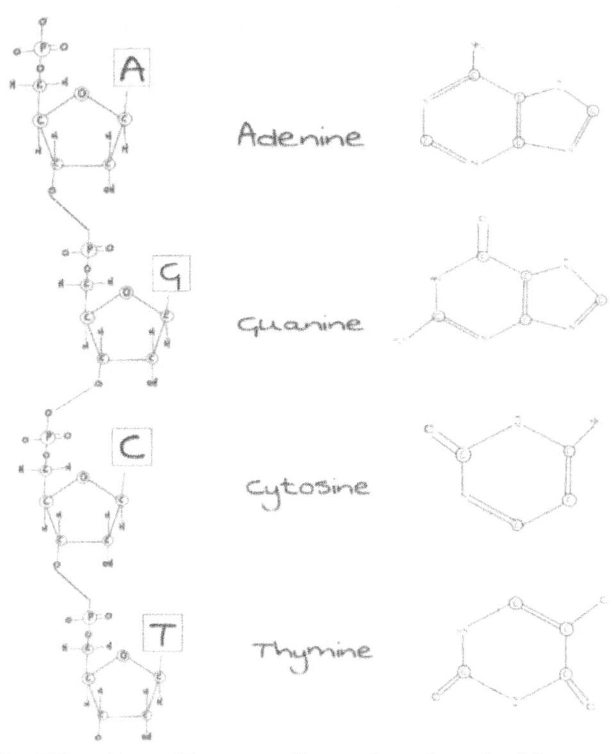

Fig. 7 Bonding of bases to ribose phosphate backbone

How is a strand of DNA stored?
We can store long threads of cotton by winding them around a cotton reel. It is easy and convenient and we know that if cotton is not stored in this way it will usually tie itself up in knots. DNA is similar, but instead of being wrapped around cotton reels, the spools are called histones. This too allows for neat packaging and storage. When the information is needed, the DNA unravels the required 'recipe' or gene and has it copied before the DNA is returned to its previous stored state.

Unlike cotton thread, the nucleotides start trying to stick to each other - a bit like unravelled sticky tape. Sticky tape sticks to any other tape it touches, but DNA only sticks together in a very specific pattern: adenine can only join to thymine, and cytosine can only join to guanine.

Consider the image that the nucleotides were either American or British electrical plugs that have either two or three pins respectively. The two-pinned adenine plug needs the American two-pinned socket of thymine, whereas the three-pinned plug of cytosine needs the three-pinned British socket of guanine. In DNA, the adenine and thymine 'plug together' and form two bonds to keep them in place, while cytosine and guanine 'plug together' with three bonds.

If DNA had only one strand then the different C and G or A and T nucleotides would join to, and form bonds with, their complementary partners and end up in a sticky mess. To stop this happening, a second strand of DNA is needed. This is a mirror image of the first strand, as shown in Fig. 8.

This becomes the recognisable double strand of DNA. Each base bonds specifically to its partner: A joins to T and C joins to G, and prevents the messiness of the DNA sticking to itself and subsequently becoming harder to unwind and copy.

More details

The bonds that form between the complementary base pairs are hydrogen bonds - where the two nucleic acid bases share either two or three hydrogen atoms between them respectively. Fig. 8 shows the details of these bonds. We can also see that the DNA is 'read' from the 5′ end in the direction of the 3′ end.

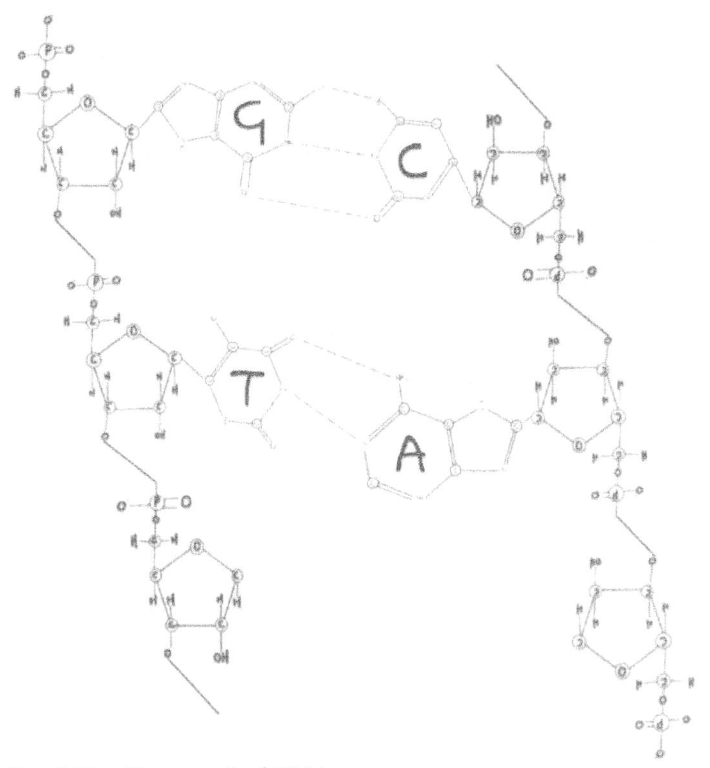

Fig. 8 Double strand of DNA

Keeping DNA information safe

Double-stranded DNA is an effective way of compressing the vast amounts of information in a cell, but nature has found a better way to store it and forms the DNA into a helix. That is, the DNA wraps itself around itself. Just as a piece of cotton held between the thumb and fingers and twisted eventually wraps around itself and collapses into a far smaller shape, so too does the DNA in eukaryotes manage a similar trick. However, DNA goes a step further and has an *even* more effective way to wrap and store the vast amount of information: it folds the twisted helix of DNA a second time and forms a double helix - that is, wrapping the first helix around itself. Old-fashioned filament bulbs used a similar shape, as seen in Fig. 9. The bulb manufacturers copied nature's idea and wrapped the tungsten filament of the light bulb once and then twice into a double helix.

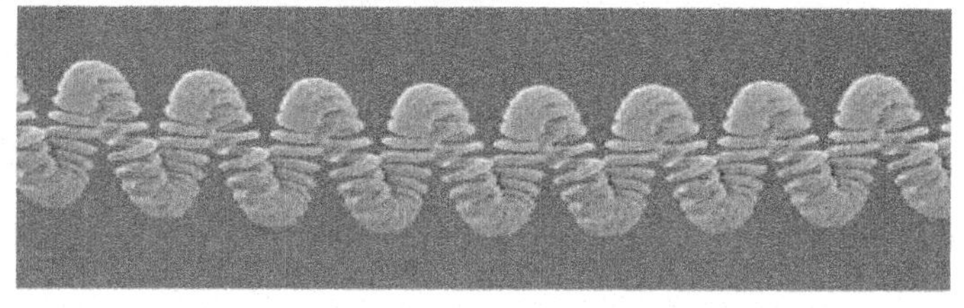

Fig. 9 A tungsten bulb filament.

Although the genes of prokaryotes and eukaryotes are fundamentally similar, they do differ in several ways. The DNA is stored in a different place, is a different size and, when it comes to expressing itself, has a different way of doing things. The eukaryotic genome is very complicated and elaborate compared to the prokaryote genome.

In eukaryotes, the storage of information is organised like a tidy office block where the information is stored neatly in filing cabinets in the basement of the building. Rather than being called the basement, the record store in eukaryotic cells is called the nucleus. This is an organelle encased in its own special membrane: the nuclear membrane. The DNA is very long but generally straight when unwound. If the DNA from one human cell were unravelled and stretched out it would be about 1.8m long! However, as the DNA is wrapped up neatly, it is compressed into a very tiny part of each and every cell.

In prokaryotes such as bacteria, the information storage is more like that of a teenager's bedroom: everything is in there somewhere, but it is nowhere as neat and tidy as in the eukaryote. There is no nucleus, and the information floats far more freely. There is generally much less information stored and, rather than being long threads, it usually forms circles and is known as circular DNA.

Size is again an issue. Generally, single-celled prokaryotes have much smaller genomes than the more complex eukaryotes, but this is not always the case, and some prokaryotes have more DNA than a eukaryote. One such prokaryote is the record-breaking bacterium *Ktedonobacter racemifer*. With 13,661,586 nucleotide base pairs (bp), this is the largest bacterium discovered so far. The eukaryote *Saccharomyces cerevisiae*, better known as brewer's yeast, has 'only' 12,156,677 bp. A human has 23,000 genes made of approximately three billion bp.

How does DNA copy itself?

A quick summary: as we saw earlier, each strand of DNA is made from sugar molecules called ribose connected by phosphate groups to other ribose sugars to form a long chain of nucleotides. To this strand a complementary copy of DNA is bound - a mirror image, as it were. This double strand is wound around proteins that act like cotton reels. It is wrapped first into a helix and then wraps around itself again to form a double helix.

Each strand of DNA is duplicated to create an identical copy. In this way, when the two strands are unwound and duplicated there will be four strands: the two original strands and two new copies. One copied strand joins to one original strand; when recombined, they form two identical double strands of DNA. This is known as semi-conservative replication and was proposed in 1958 by Meselson and Stahl using *Escherichia coli* (*E. coli*) bacteria.

DNA is circular in prokaryotes but linear in eukaryotes. As a result, the mechanisms of duplication are slightly different. As we are human, we will look at what happens when our eukaryotic DNA wants to copy itself. First it needs to be unwound and the two strands 'unzipped' from each other. The 'unzipping' enzyme is called DNA helicase. This enzyme breaks the bonds that had formed to link up the nucleotide base pairs. In effect, this is like taking the plugs out of their sockets.

Once unwound, the process of copying the DNA needs another helping hand, as replication can't get started on its own. A bit like a child in a maths class, once they have been shown the first example they can work through the whole worksheet without much input. 'Priming' is needed to get the process going and is done by another enzyme: RNA primase. This creates a 'sticky end' to allow a builder enzyme to be able to add nucleotide bases on and so lengthen the chain of new DNA. In the cell, the builder enzyme is DNA polymerase.

The builder will fall off the DNA if it isn't held in place. The safety harness that keeps it attached to the DNA is called a sliding clamp. This clamp, as the name suggests, slides along the strand of DNA as new nucleotides are added to the newly produced strand. The clamp structure is shown in Fig. 10. It takes energy to take the clamp out of the box and attach it to the DNA strand. Once in place, the 'clamp loader' can sit back and let the DNA replicate itself. The energy needed for this comes from a very special molecule called ATP, which we will meet in Chapter 4.

The strands of DNA try to curl themselves back into a helix shape; copying can only happen if they are flat and straight, so a helper is needed. Single-stranded binding proteins come to the rescue and hold each strand straight and smooth and remove any kinks or turns. The duplication process is allowed to proceed.

More details

At the heart of the replication process is DNA helicase. This enzyme splits the hydrogen bonds of the base pairs to create two single strands of DNA; these are held apart by single-stranded binding proteins while copying occurs. DNA gyrase or topoisomerase is the enzyme that prevents the knotting or supercoiling of the rest of the DNA molecule.

As we saw earlier, the carbon ring is labelled like a clock so that each carbon atom can be identified. Starting at the fifth carbon - known as the 5 prime or 5′ end - replication happens in the direction towards the third carbon – or 3′ end. Therefore, the new piece of DNA is made in the direction of 5′ -> 3′.

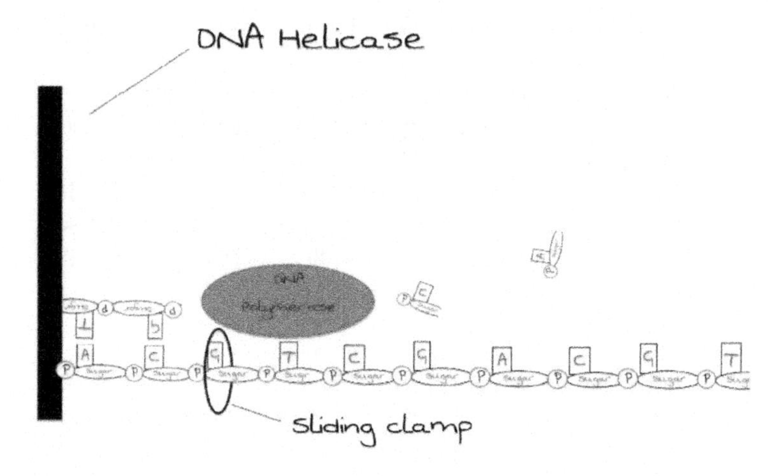

Fig. 10 The sliding clamp holds the DNA polymerase in place as the DNA helicase holds the two strands of DNA apart

Priming needs to be done before replication can begin. DNA primase creates an RNA primer - a molecule 8-10 nucleotides in length. The primer binds to the parent strand of DNA and new nucleotide bases are attached to the third carbon end of the primer. Joining a new nucleotide to the oxygen atom from a hydroxyl (-OH) group means a hydrogen atom is lost in the process.

Priming is needed as DNA polymerase can only add a nucleotide to a previous one. In eukaryotes, several DNA polymerases are involved in the replication of DNA. DNA polymerase III adds nucleotides to the strand of DNA. However, in prokaryotic bacteria, such as *E. coli*, the enzyme DNA polymerase II is able to 'proofread' the new DNA strands while DNA polymerases IV and V are involved in DNA repair.

The sliding clamp assembly is a ring-shaped protein complex formed from two parts: the sliding clamp and the clamp loader. ATP is used to take the clamp loader and bind the clamp and DNA polymerase onto the strand of DNA.

Finally, another DNA polymerase - DNA polymerase I in eukaryotes - replaces the RNA primer with a segment of DNA.

Leading and lagging DNA strands

Like walking down a one-way street, DNA can only be copied in one direction. When DNA is replicated and the two strands unzipped, one strand becomes the leading strand and the other the lagging strand. The leading strand can be copied easily as one long continuous strand. Unfortunately, there is a problem with trying to copy the lagging strand: it can't be copied against the flow and can't go in the direction it needs to. However, DNA has found a way around this. DNA is copied in short segments, still in the 5'->3' direction, and then linked together to form a single long strand, as in Fig. 11.

When the process is complete, there are the two original DNA strands - the lagging and the leading strands - and one copy of each of them. This gives a total of four strands. Each pair - one new strand and one original strand - are wrapped back together, producing two identically paired strands of DNA, as presented in Fig. 12. We have now successfully copied our DNA!

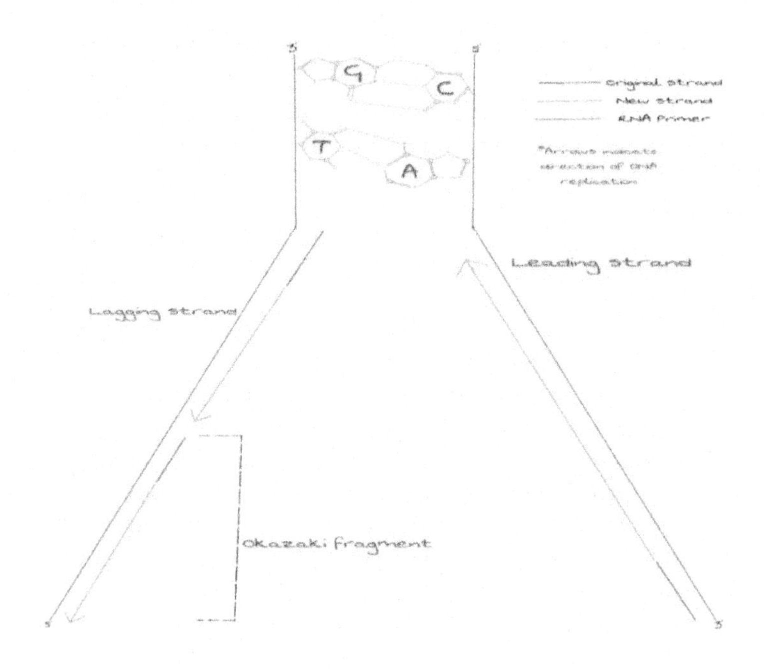

Fig. 11 Extension of the DNA sequence

More details

The little pieces of DNA copied from the lagging strand are called Okazaki fragments, after their discoverers in the 1960s. These short segments of DNA are bound together by an enzyme called DNA ligase into one continuous chain or the second daughter strand.

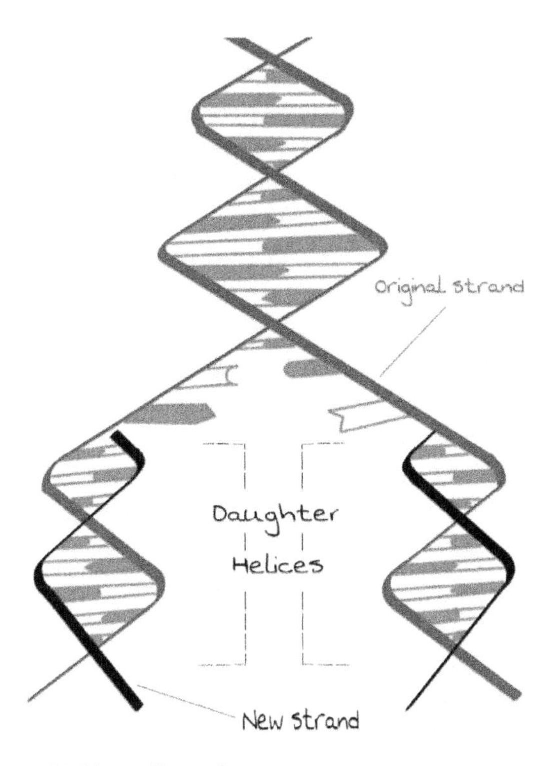

Fig. 12 DNA replicated

Using the information in a cell

We have seen how information is stored and copied within a cell. The overall goal for DNA is to be turned into proteins. The single-celled prokaryotes have their DNA distributed within the cytoplasm, whereas the eukaryotes have it neatly packaged into the nucleus. How does the information get to the factory floor where it will guide the manufacture of these proteins? This process takes a few steps and along the way results in the production of another key polymer: ribonucleic acid (RNA).

RNA

When the information stored in the basement of a factory is required elsewhere, it can be photocopied and sent to the different departments. These copies of the blueprints will guide the machinery into making the car parts or kitchen utensils or whatever the factory is making. Importantly, the original blueprints never leave the basement. In a cell, the photocopies of the original information are known as messenger RNA (mRNA).

If you wanted a chocolate cake recipe from a book in the library, you could copy the pages on which the ingredients and method are written to take away with you. Unfortunately, the photocopied pages are not as robust as the book; they are more likely to end up with chocolate spillages and butter stains and as a result the copy won't last as long as the original. The same is true for mRNA. The method of copying DNA into mRNA is called transcription. It *transcribes* the original information from the DNA into a copy as mRNA. This can be likened to monks sitting in their cells in the monastery writing out the Bible before the introduction of printing presses.

The nuclear membrane surrounding the DNA has holes or pores in it. These act like letterboxes. Just as you can't pass a filing cabinet through a letterbox in a front door, DNA is too big and bulky to leave the nucleus. However, you can pass photocopied sheets of paper through, and this is how information leaves the nucleus as mRNA. In reality, the mRNA is a smaller and straighter polymer of nucleotides than the relatively big and complex wrapped shape of DNA. As only a small part of the total information is needed at any one time, mRNA is considerably shorter than DNA. Another way to describe this is that RNA is a relatively short polynucleotide chain. Unlike DNA, mRNA doesn't need any proteins to wrap itself around. The mRNA photocopies are not exactly the same as the original. When photocopying, the paper may be a different colour or a different size to the original information being copied. In copying DNA to mRNA there is a crucial difference: one of the nucleotide base pairs is changed. In DNA, adenine was binding to thymine and cytosine to guanine. In mRNA, thymine is not used; it is substituted by uracil, represented by the letter U. So in mRNA, C still binds to G but A now binds to U.

Even after the mRNA has been generated from the DNA in the nucleus, it is still not quite ready for distribution; it needs some finishing touches. We may laminate the photocopies we have made of our chocolate cake recipe or put them in a plastic binder to help them last a bit longer. The mRNA faces danger in the cytoplasm if it leaves the nucleus without protection. With this in mind, the mRNA plays safe and has a 'cap' and 'polyA tail' added to either end. This is to try to stop it being broken down and degraded by enzymes once in the cytoplasm. The cap is added to the 5′ end and is a modified GTP molecule (a molecule a bit like the ATP we'll meet in Chapter 4). The 'polyA tail' means there are up to 150 adenine units in a row at the 3′ end.

The factories

Information has been photocopied, has left the filing department and found its way to the factory floor. How is this turned into the rolling pins or engine block the factory is making? The cell doesn't make car parts but does make proteins. Proteins are amazing molecules that do many different jobs. Some act as bricks and cement or steel girders in building the cell walls; others act as military escorts to take small molecules through the body unimpeded. Some speed up or slow down chemical reactions, in effect working like the temperature control on an oven so reactions are neither too fast nor too slow. Others act like emails - sending messages around the body to start other cells working correctly. Some act with the body's own army or white cells to protect against invaders (these are the antibodies we will meet in Chapter 6).

How are proteins made?

Proteins are made from smaller building blocks called amino acids. These are stuck together like rings in a paper chain to form a long chain. Although more than 500 amino acids have been found, only 22 are needed by humans to produce a huge variety of proteins. Of these 22, only nine are considered 'essential' amino acids. This means they can only be sourced through diet; the body can make the others itself.

Before we can make a protein, we need to meet another molecule called transfer RNA (or tRNA). This molecule holds an amino acid with one hand and, with the other hand, searches for someone to help their amino acid to become part of a protein. The help comes in the form of a protein complex (more than one protein working together) called a ribosome. A ribosome is able to 'stick' the amino acids together by transferring it from the tRNA onto the lengthening chain of amino acids. To know which order to combine amino acids together, a ribosome can 'read' the instructions given to it by the mRNA. In our box in Chapter 1, the necklace maker is the ribosome.

Making a protein using paper chains of amino acids

Another way to think of a protein is with a paper chain. For this practical, you will need:

- A hat labelled 'ribosome'.
- A group of willing people to act as transfer RNA molecules.
- Strips of different colour papers, each colour representing an amino acid.
- Sticky tape to bind the amino acids together.
- Pieces of plain paper and a pen that will become the messenger RNA.

The rest of the group (bar yourself!) write their names on the strips of plain paper. In return for their name, they receive a strip of coloured paper. All the named strips of paper are laid out across the tabletop forming a long chain - this is the information contained within a mRNA molecule. The 'ribosome' (the person wearing the 'ribosome hat') calls the first name from the strips of plain paper. The named person or 'transfer RNA' forms their coloured paper/ 'amino acid' into a loop and, with one hand holding the coloured loop, holds a hand of the ribosome. When done, the second name from the mRNA is read out. This 'transfer RNA' holds a hand of the ribosome and forms a loop with their coloured 'amino acid' inside the first loop creating the beginnings of a chain. As this is happening, the first tRNA is let go by the ribosome and released back into the nucleus to find another amino acid to grab hold of and bind to. The ribosome still has hold of the second tRNA and the third name on the mRNA is read. The corresponding tRNA is 'called' to the ribosome. As the amino acid is joined to the lengthening chain, the second transfer RNA is let go. This continues until all the mRNA is read and a long paper chain of amino acids has been completed and represents a protein!

The beads (or amino acids) are threaded onto a cord to produce a chain of beads or protein.

How does the ribosome know which amino acid is next?

In the practical example above, the 'ribosome' was able to read the name of the next person in order to make sure the right person came forward with their 'amino acid'. In a cell, the ribosome reads the mRNA. Earlier, we found that the long string of 'letters' all in a row form DNA; when in mRNA, the chain is made of adenine, cytosine, guanine and uracil. If each letter were read individually (A, C, G or U), there are four possible outcomes. If the bases were read in pairs, there are 16 variations (or 4 x 4). They are: AA, AC, AG, AU, CA, CC, CG, CU, GA, GC, GG, GU and UA, UC, UG and UU. This is a fourfold increase in the number of possibilities. Bases are actually read as a group of three. This extends the number by a further order of power to a total of 64 (or 4 x 4 x 4). This means that every three nucleotide bases is read as a single unit, known as a **triplet** or a **codon**; for example, AAA or AUG or CUA or GAC. Each of these triplet codes is for an amino acid; this is known as the **universal code**.

In Fig. 13 we can see the codon and the corresponding amino acid. Some amino acids have several possible codons. Imagine you want to ring a friend, but are missing the last digit of their phone number. You could keep trying different combinations: after some wrong numbers and a bit of luck, you might get through to them. The codon for, say, glycine (Gly) means it doesn't matter which of the numbers you use for your friend's final digit, you will still get through first time. Any codon that starts 'GG...' will be glycine; even if a mistake has taken place in the copying of the DNA into mRNA, the amino acid will still be glycine. This is known as the degeneracy of the genetic code. Some amino acids have more than one codon specifying for them during protein synthesis.

<table>
<tr><th colspan="8">Second letter</th></tr>
<tr><td rowspan="17">First letter</td><td></td><td>U</td><td>C</td><td>A</td><td>G</td><td></td><td rowspan="17">Third letter</td></tr>
<tr><td rowspan="4">U</td><td rowspan="2">UUU} Phe
UUC}</td><td rowspan="4">UCU}
UCC} Ser
UCA}
UCG}</td><td rowspan="2">UAU} Tyr
UAC}</td><td rowspan="2">UGU} Cys
UGC}</td><td>U</td></tr>
<tr><td>C</td></tr>
<tr><td rowspan="2">UUA} Leu
UUG}</td><td>UAA Stop</td><td>UGA Stop</td><td>A</td></tr>
<tr><td>UAG Stop</td><td>UGG Trp</td><td>G</td></tr>
<tr><td rowspan="4">C</td><td rowspan="4">CUU}
CUC} Leu
CUA}
CUG}</td><td rowspan="4">CCU}
CCC} Pro
CCA}
CCG}</td><td rowspan="2">CAU} His
CAC}</td><td rowspan="4">CGU}
CGC} Arg
CGA}
CGG}</td><td>U</td></tr>
<tr><td>C</td></tr>
<tr><td rowspan="2">CAA} Gln
CAG}</td><td>A</td></tr>
<tr><td>G</td></tr>
<tr><td rowspan="4">A</td><td rowspan="3">AUU}
AUC} Ile
AUA}</td><td rowspan="4">ACU}
ACC} Thr
ACA}
ACG}</td><td rowspan="2">AAU} Asn
AAC}</td><td rowspan="2">AGU} Ser
AGC}</td><td>U</td></tr>
<tr><td>C</td></tr>
<tr><td rowspan="2">AAA} Lys
AAG}</td><td rowspan="2">AGA} Arg
AGG}</td><td>A</td></tr>
<tr><td>AUG Met</td><td>G</td></tr>
<tr><td rowspan="4">G</td><td rowspan="4">GUU}
GUC} Val
GUA}
GUG}</td><td rowspan="4">GCU}
GCC} Ala
GCA}
GCG}</td><td rowspan="2">GAU} Asp
GAC}</td><td rowspan="4">GGU}
GGC} Gly
GGA}
GGG}</td><td>U</td></tr>
<tr><td>C</td></tr>
<tr><td rowspan="2">GAA} Glu
GAG}</td><td>A</td></tr>
<tr><td>G</td></tr>
</table>

Fig. 13 Universal genetic code (see end of chapter for abbreviations)

So we have our messenger RNA with three base pairs to be read together. Part of the ribosome reads this and knows what sort of amino acid is needed next. Swirling around it in the cytoplasm are hundreds of tRNA-amino acid complexes waiting their turn to be next. It's a bit like Cinderella and the glass slipper - only one will be the correct fit. If in the previous practical, the wrong person or tRNA-amino acid complex came up, the ribosome would turn them away because it is following strict instructions written in the mRNA.

This whole process is called translation because it *translates* the information from the RNA into proteins.

The ribosome forms bonds between the amino acids. Instead of using sticky tape, the bonds are called peptide bonds. The chain of amino acids forming the proteins is called the primary structure, but this in itself is not useful to the cell. The protein needs to be folded: just as steel bars at a steel works are pressed and folded and bent and riveted together to form the finished product, the same happens in cells. During the production of the primary protein structure, the chain is held straight by supportive molecules called heat shock proteins.

So, we have DNA, which is a string of nucleotide bases on a sugar phosphate backbone. This is transcribed into mRNA in the nucleus, which travels to the cytoplasm to be translated into a string of amino acids known as a protein.

How does the ribosome work?

The ribosome is made of two parts or subunits. A bit like a kitchen mixer - made from the mixing bowl and the mixing machinery - neither will work without the other. The two parts of the ribosome come together and read the mRNA 'recipe'. On the work surface are the ingredients - the tRNA-amino acid complexes waiting for their turn to be added to the bowl. The ribosomes in eukaryotes and prokaryotes are different sizes, the former having a 60S large subunit and 40S smaller subunit, the latter 50S and 30S respectively. The 'S' stands for Svedberg, a unit used to measure how fast molecules move in a centrifuge. This gives an indication of the size, shape and density of the molecule.

The ribosome has three binding sites labelled A, P and E. One is for arrivals of a new piece of tRNA (A site = aminoacyl); one is for processing the tRNA (P site = peptidyl); and one is for exiting (E site) the ribosome, as shown in Fig. 14. The mRNA is read like a ticker tape running between the large and small subunits of the ribosome.

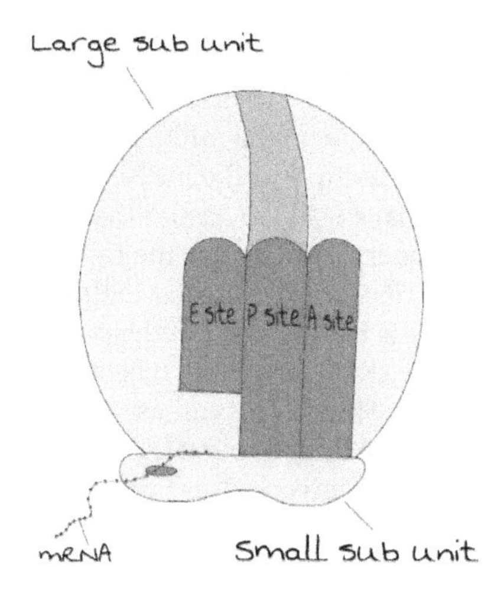

Fig. 14 Binding sites of a ribosome

As the ribosomes are different between prokaryotes and eukaryotes, some antibiotics can be made to specifically target this cellular machinery. For example, tetracycline specifically blocks the A site of bacterial ribosomes.

How does the ribosome know where to start reading? Or, 'ho wdoe sth eribosom ekno wwher et ostar treading?'

This is the same sentence, but the last letter of each word is transferred to the beginning of the following word. This makes the message unintelligible. By knowing where to start reading a sentence (the capital letter) and where to stop (the full stop), we can read and understand complex sentences. Ribosomes do the same thing by 'setting the reading frame'; i.e. knowing where to start reading the message from. If the reading frame is incorrectly set, then the message becomes unintelligible.

More details

The tRNA is shaped like a clover leaf, with three leaves and a stalk. The stalk attaches to the amino acids. On the opposite 'leaf', there is a triplet of bases that will be the reverse of the message on the mRNA. This is called the 'anti-codon'. For example, if the mRNA has AUC then the tRNA must have UAG to ensure the fit is as snug as Cinderella's slipper. This ensures the right tRNA is attached by the ribosome according to the mRNA, so the right amino acid is bonded next in the protein chain. tRNA is made by tRNA synthase, which has a proofreading site. The tRNA synthase checks the correct amino acid has been added to the tRNA. If an incorrect one has been added, it recognises the error and the amino acid is removed.

When the smaller subunit of the ribosome has bound to the mRNA at the ribosome binding site, there is a series of base codes (usually AGGAGG) to tell the ribosome to get ready (a bit like a paragraph break in a page of text). After these base codes, there is a small gap or space before the start codon (akin to the capital letter in our sentence) - this is usually AUG. This code is for the amino acid methionine and initiates the production of the protein. Once the methionine is in place, the large ribosomal subunit joins in. Amino acid methionine is in the P site, leaving the A site unoccupied. The A site 'reads' the mRNA and now knows which codon it needs to find on the 'leaf' of the tRNA to ensure it gets the right amino acid.

The bonds are formed between the amino acid on the tRNA at the P site and the A site. When this is complete, the A site wants to move on to the next codon. To do this, the conveyor belt moves along. The tRNA now in the E site is no longer attached to an amino acid, having had the bond broken; the amino acid then joined the amino acid in the A site. This leaves in search of a new tRNA synthase and amino acid to bond with. This in turn frees the A site for the next triplet to be read and tRNA to be found. This is the elongation phase of protein synthesis.

The termination phase is when the protein is complete and all the mRNA has been read. The end of the book is reached, so it is time to put it down and start reading the next book. The final full stop of the mRNA code is either UAA or UAG. When this appears in the A position of the ribosome, a release factor binds to the site instead of a tRNA-amino acid complex. The ribosome moves this release factor into the middle P position where the synthesis of the protein stops and the end is 'sealed' by adding a hydroxyl group (OH) to the last amino acid. This stops any further amino acids from binding to the chain.

Summary

DNA is formed from a sugar phosphate backbone onto which are added nucleotide bases. The two mirror image copies of the DNA strands combine and then wrap and wrap again to form the double helix storage system. DNA replicates itself using a series of enzymes to help create two new identical strands during semi-conservative replication.

Universal code explained:
Leu - Leucine; Phe - Phenylalanine; Ile - Isoleucine; Met - Methionine; Val - Valine; Ser - Serine; Pro - Proline; Thr - Threonine; Ala - Alanine; Tyr - Tyrosine; His - Histidine; Gln - Glutamine; Asn - Asparagine; Lys - Lysine; Asp - Aspartic acid; Glu Glutamic acid; Cys - Cysteine; Trp - Tryptophan; Arg - Arginine

Chapter 3 - Genes working right and wrong

The laws of genetics apply even if you refuse to learn them.

- Allison Plowden

Genes do amazing things. Just a few letters of DNA in the right order make everything around us - from plants and animals to bacteria via individual proteins. Genes respond to the environment around them by being switched on and off. Unfortunately, this sensitive equipment sometimes makes mistakes - either when it is transcribed or when it is translated - and this can have profound effects for the organism. Sometimes the organism gains from the mistakes, but sometimes it suffers.

Genes working right

When genes are turned 'on' it means that they are being 'expressed'. Gene expression signifies that the DNA is being transcribed into RNA, and this RNA is being translated into proteins, as we saw in Chapter 2. Some genes are always 'on' and their proteins are constantly being produced. Other genes are not always being expressed and are 'off'. Whether or not a gene is expressed is controlled by promoters or inhibitors. These are molecules or chemicals that regulate the switch. In the home, the flicking of the light switch into the on position allows the electricity to flow and the light turns on. If the switch is not in the 'on' position, there is no light - or, in the cell's case, no proteins or final product. In cells, inhibitors and promoters work in a similar way. Some genes are only expressed in certain conditions. The lights in a house are only turned on at night when they are needed but not during the day when it is light outside.

Inducible genes are those that can be turned on when certain conditions are met. For a light to be turned on, it needs both to be dark outside and for the light to be switched on.

The discovery that gene expression could change was demonstrated by examining the *E. coli* bacterium and its response to different foods. In experiments done in the 1960s by Nobel Prize winners Jacob and Monod, *E. coli* was put into a mixture of lactose and glucose in water. Glucose is sugar and bacteria love glucose in the same way nine-year-old kids love chocolate cake at a birthday party. Given the choice, both children and bacteria would eat as much glucose as they could! *E. coli* reproduces rapidly in these ideal conditions. This is shown in Fig. 1 by the first steep climb on the graph. As we will see in Chapter 4, the energy from glucose is stored as ATP. This energy is available to help a cell to grow quickly and divide rapidly. Lactose is a disaccharide and is made of two different sugar molecules bonded tightly together. The cell cannot use this as a food source without breaking the bonds apart.

What happens when the food or glucose runs out? The tantrums of nine-year-olds when they realise that there are only celery sticks and slices of pepper left are nearly as devastating as an *E. coli* running out of glucose. When *E. coli* runs out of glucose it stops replicating and growth plateaus, as there is no more suitable fuel they can use. However, returning to Fig. 1, there is a second period of rapid growth and division after the horizontal plateau. Why is this and how does it occur?

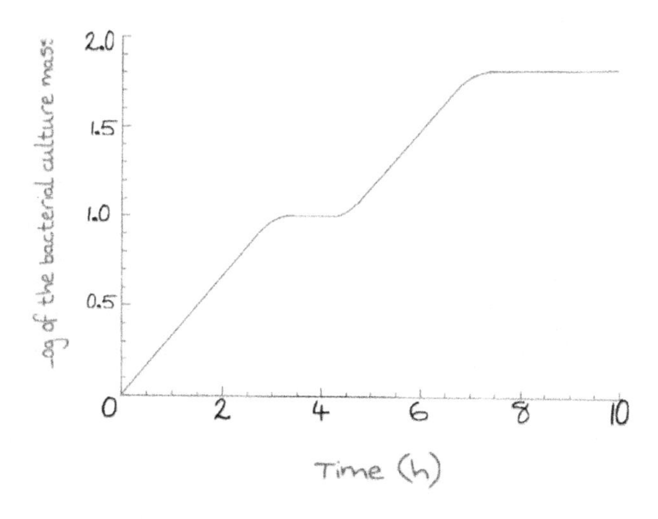

Fig. 1 Growth of *E. coli* in a mixed glucose and lactose solution

How does E. coli use a different food source?

In the mix of glucose and lactose, *E. coli* doesn't want to waste any time, energy or effort making the proteins and enzymes it would need to digest lactose. It is too busy using the glucose and enjoying the table of sweet party foods too much. The cell membrane is very good at keeping lactose out and only letting glucose into the cell; it is said to be 'impermeable to lactose'.

Within the DNA of *E. coli* lies a group of genes that can be turned on when it needs to make use of a different food source. When expressed, this gene is transcribed into mRNA, then translated into proteins that use the new food to make energy and allow growth to resume. In this case, the alternate food source is lactose and the gene controlling operations is the 'lac operon'. An operon is 'a group of genes under the control of a single promoter'. Lactose is made of a combination of half glucose and half galactose. To allow the cell to make use of lactose, a lot of buttons and levers need to be pushed and pulled in the right order. In reality, the lac operon needs to be expressed and start producing the required enzymes.

The lac operon is made of three genes in a row:

• lacZ is the gene for β-galactosidase. This enzyme breaks lactose down into the two sugar molecules of glucose and galactose.

• lacY is the gene for β-galactoside permease, which opens channels across the cell membrane to let lactose in.

• lacA is the gene for thiogalactoside transacetylase - function unknown.

When these genes are working and being expressed, the cell membrane lets lactose into the cell, where it is broken down into glucose and galactose. Expressing a gene takes time, energy and resources, which the cell doesn't want to expend unless absolutely necessary.

We know that the lac operon isn't usually 'on'; this on/off switch is controlled by other genes that precede it. The lac operon looks a bit more like this:

• lacI is the inhibitory gene or the 'off switch'.

• the next gene is the CAP site or 'catabolite activator protein' site.

• the last gene before lacZ is the 'operator site' or 'on' switch.

In normal conditions, lacI is always 'on' and is being expressed. It produces a molecule that blocks or inhibits the rest of the operon from working. This repressor molecule is a string of mRNA that binds to the DNA at the operator site. It blocks expression or transcription of lacZ, lacy and lacA. In effect, the 'on' switch can't be turned on.

When glucose is plentiful, the cell uses it to generate ATP. The enzymes that allow lactose to cross the cell membrane or break it down into simpler sugars are not even being made when glucose is available. But when the children have devoured the chocolate cake and the *E. coli* have used all the glucose, what happens next? Without glucose, ATP can't be produced; without energy, the cell machinery grinds to a halt. However, the cell can store energy in a different way, using a molecule called cyclical AMP or cAMP. This is not as good as ATP for energy but is better than nothing. cAMP is able to bind to the CAP site, and this is an 'on' switch for the lac operon. Unfortunately, cAMP on its own cannot start the expression of the lac operon because the operator site is still being blocked by the repressor mRNA.

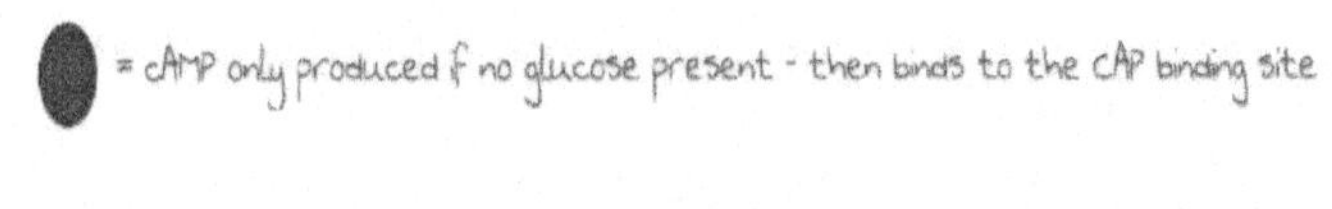

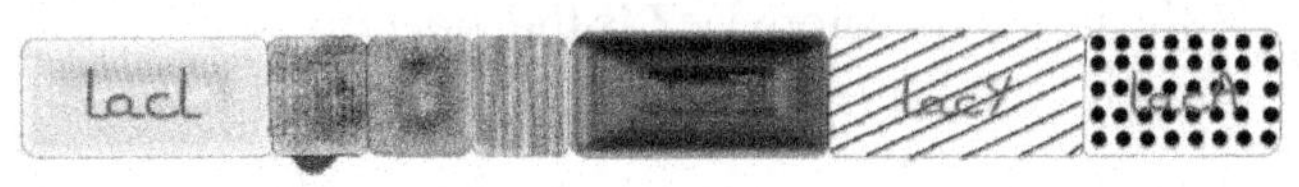

For the lac operon to be expressed, the blocker mRNA molecule needs to be taken out and the CAP 'on' switch needs to be flicked. The on switch promoter gene is turned on by cAMP, but the blocking mRNA molecule also needs to be stopped. For this, a secret exists: the blocking mRNA can itself be blocked. The magical molecule that does this is lactose.

But how does lactose get into the cell? The cell membrane was doing an excellent job at keeping it out. But a single lactose molecule using the stealth of a ninja and the cunning of a fox can and does get into the cell. The lactose molecule finds a weakness - a slight leakiness in the cell wall - and breaks in. This one molecule can take out the blocker molecule and, if there is no glucose present so cAMP is present on the CAP site, the lac operon can start producing mRNA and in turn create the enzymes needed.

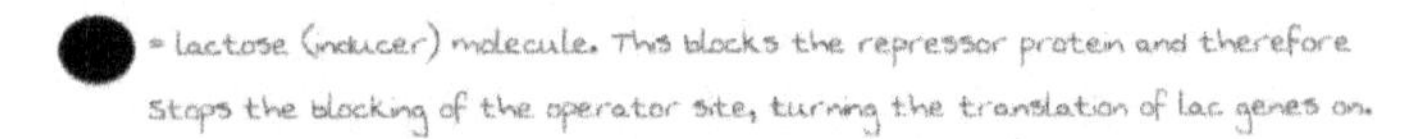

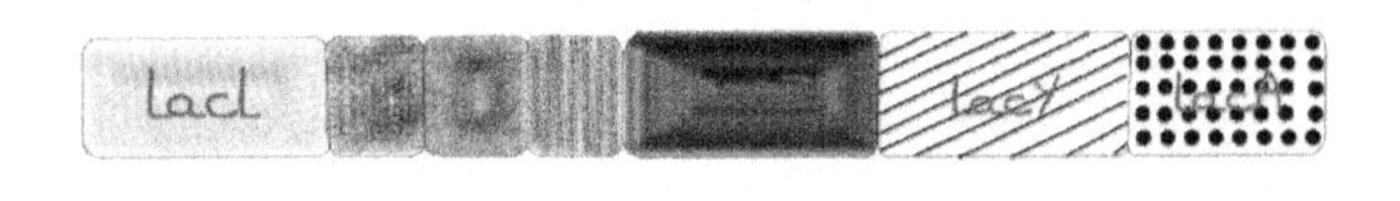

Therefore, to metabolise lactose, the cell needs a lactose molecule to block lacI mRNA and cAMP to bind to the CAP binding site.

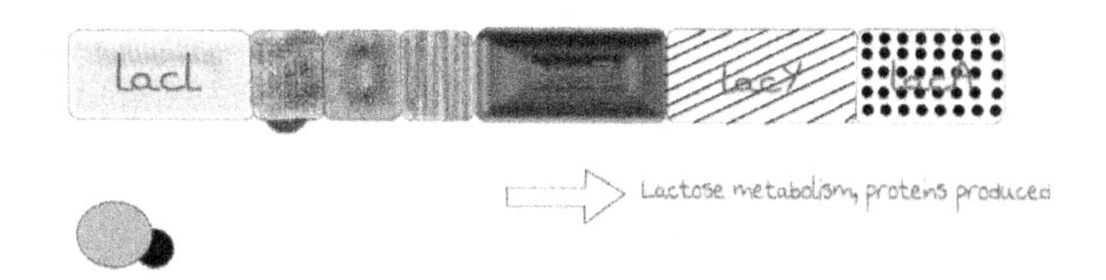

Now the machinery springs to life. The mRNA starts churning out the enzymes. The role of one of the β-galactoside permease enzymes is to make the cell membrane more and more leaky, or permeable to lactose. It travels to the cell membrane to let in more lactose and the more lactose that enters the cell, the more lacI mRNA that is blocked.

The second enzyme is beta-galactosidase. This puts the lactose to work by breaking it down into glucose and galactose. The new-found glucose is used by the cell to make ATP again.

Only when the conditions are right - when no glucose is present and a single molecule of lactose is present - can the cell make the proteins needed to have more lactose enter the cell, be broken down and be used.

This allows the *E. coli* to start growing and dividing again, and normal service is resumed. This is the second portion of rapid growth on the graph in Fig. 1.

Genes going wrong

The lac operon is an example of genes that work only when they are needed; i.e. when the environmental conditions are right. The cell does not waste resources and energy making proteins and enzymes or mRNA without good reason. If the right sequence of nucleotides is present in DNA and these are successfully converted to a protein, then the cell functions normally. But what could go wrong with a gene?

What are mutations?

Mutations are changes to the DNA. These errors can occur during the copying or replication of DNA, the change or transcription into mRNA, or during translation or production of the protein. The most important changes or mutations are those that affect the final protein, as the protein is the functional part of the cell. The effects of mutations can range from no effect to major effects on the DNA and this in turn can lead to no effect through to catastrophic effects on the structure and function of the protein.

How do mutations occur?

If an error occurs when DNA is copied into the daughter strands, it may not be noticed. There are proof-checking mechanisms during the copying process – in the same way a book is proofread for mistakes, so it is for DNA.

A missense mutation occurs when one letter in the DNA code is substituted for another. Although only a single letter has been changed out of a code of many thousands, this single base mutation can have important effects, as it affects the codon that codes for a specific amino acid. The effect becomes apparent when the wrong information is coded into the mRNA that is sent to the ribosomes to be turned into proteins.

The effect of a missense mutation is seen in the difference between the genes of someone with wet waxy ear canals compared to those with dry ear canals. A study in 2002 in Japan showed that this was due to a single nucleotide base change on the ABCC11 gene. In wet ear wax, there is a G at site 538; in dry wax ears there is an A. This substitution shifts the amino acid from glycine to arginine and has a knock-on effect on the shape of the protein produced. Fascinatingly, some regions of the world, including Asia, produce more dry ear canals. In Europe and Africa, the wet ear wax mutation predominates. Is there an evolutionary reason for favouring one type of ear wax over another? Are humans more or less likely to choose a partner and parent to their children based upon the type of ear wax they have? I suggest this would be a very superficial reason for discriminating for or against a future partner, but what if the gene for wet ear wax also affects sweat production? As sweating is a method for cooling the body in hot weather, if someone sweated more, they may cool themselves faster. This could give them an evolutionary advantage, for example, when cooling down after hunting prey, or the ability to work longer as a farmer or need fewer rest breaks. Alternatively, when we examine evolution in Chapter 5, this may be regarded as a neutral genetic change.

Of greater significance is a condition called sickle cell anaemia. This is caused by a single base substitution in the gene for beta-haemoglobin. Beta-haemoglobin is a protein used in the manufacture of red blood cells. Normal red blood cells in adults are made from four protein molecules: two alpha chains and two beta chains. These proteins fold themselves around iron atoms. Oxygen binds to the iron atoms in the red blood cells and is transported around the body to the muscle cells. Here, the iron releases the oxygen to enter the mitochondria and is used in the generation of the energy that is stored as ATP (see Chapter 4). In other animals, such as crabs, copper atoms are used instead of iron; as a result, they have blue-coloured blood. In humans, the genes for the two haemoglobin proteins are found on chromosome 11.

In sickle cell anaemia, there is a single change in the base pair on the sixth codon for the beta protein chain. This changes the amino acid from glutamic acid to valine. This is enough to not only change the protein but to affect the shape of the whole cell from a squashy disc to a sickle shape. Normally, blood cells are able to pass through the small blood vessels (capillaries) by squashing themselves up and passing through the tight, narrow spaces without being trapped. Unfortunately, in sickle cell anaemia, the sickle-shaped blood cell cannot compress itself and can become wedged in the blood vessels. In cold weather, this is worse as the capillaries contract and become even smaller than usual. When the blood cells can't flow because of the blocked abnormally shaped cell, this causes a sickle cell crisis and is very painful. However, in hot climates, where there is a threat from malaria, it is thought that sickle-shaped red blood cells offer a degree of protection. Malaria is discussed in detail in Chapter 8, but briefly, the parasite Plasmodium is transmitted by the female mosquito to the host. Once in the host, the red blood cells are ruptured open, which contributes to the illness and potentially death. Sickle-shaped cells are less able to be broken up and remain intact.

A silent mutation is a mutation that has no effect on the final protein. If we refer back to the universal code in the previous chapter and compare a codon of GCU, which codes for alanine but change the final base pair to a C so the codon reads GCC, this would still code for alanine.

As a result, the subsequent amino acid sequence and functional protein structure would be unchanged. If a mutation occurs to a gene that is not usually expressed, then the change may not be known about for a very long time. Think of a book in a library that is only borrowed once every ten years and has a mistake buried at the bottom of page 692. It would be very unlikely that the error would be noticed.

If the codon altered and ended up being a UAA, UAG or UGA, protein synthesis grinds to a halt, as these are known as stop signals or 'stop codons'. This type of mutation is known as a nonsense mutation.

A point mutation is when an extra nucleotide base is added (insertion) or removed (deletion), leading to a frame shift or change to the reading frame. Changes to the DNA can also occur by inserting or deleting whole sections of a gene. This is likely to have a profound effect on the final protein. As we saw in the last chapter, the setting of the reading frame is vital to understand the message. If the DNA is misread, the effect on the final protein will vary. For example, if a gene is 1,000 bases long and it is in the position of the third base pair that a nucleotide base is added or removed, the change in reading frame will have a knock-on effect on the entire gene. However, if it were the 997th gene position where the change occurred, the effect may be far less pronounced, as only the last one or two amino acids would be affected. In the case of protein production, the folding post-production is vital, as this determines the function of the protein. If this is disrupted by changing the amino acid sequence, the protein would no longer work. However, the mutation may give the cell an advantage - for example, by producing a change in the enzyme shape, which allows the cell to make use of a different or novel food source and thereby allows it to grow and replicate faster than its rivals.

It can't be said that all mutations are good or all mutations are bad. Some bring positive benefits; some, such as sickle cell anaemia, can be a double-edged sword.

Wasted DNA?

So far, so good. A long string of amino acids has been joined to make a protein as DNA was transcribed into RNA and then translated into the protein. Easy peasy.... Except it isn't quite that neat in nature.

Single-celled prokaryotes such as bacteria do use the system described. If a prokaryotic cell needs to make several proteins - for example a burger, bun and salad - it will put the 'recipes' or genes next to each other so it can make all three simultaneously, as we saw with the enzymes from the lac operon. This seems a sensible way to arrange things. At home, people often organise themselves in a similar way. All the cleaning products are together, all the DIY products, all the kitchen and foodstuffs. It would be difficult to paint a wall in the house if the paint were stored in the bathroom, the paintbrushes in the kitchen, the ladders in the bedroom and dustsheets in the sitting room. In prokaryotes, the genes are stored so that their genomes have a compact organisation. So, in an *E. coli* cell, 89% of the DNA codes for proteins. There is relatively little wasted space.

Unfortunately, in multicellular eukaryotes, this is not how things are organised. The bits we need - the paint, brushes and ladder - are still coded for by genes, but are stored all over the place. In genetic terms, exons are genes that 'do' something, i.e. code for a protein. This is the coding region of the DNA. The sequences of nucleotides that are not needed and serve as gaps in the useful DNA are called introns; they have been described as 'non-coding DNA'. Just because the DNA doesn't code for a protein doesn't mean it's not useful. Some non-coding DNA is also responsible for the expression of adjacent genes, as we saw in the last chapter. In humans, it is estimated that a whopping 98.5% of the human genome is made up of non-coding DNA!

The non-coding DNA almost feels as if random paragraphs from one book are being put into another book. It wouldn't make any sense to the story being told in the first book. Some sections of DNA are repeated several times - sometimes with small changes to a few letters here and there - while others are remnants of previous evolutionary history.

However, even stranger is that the gene for some proteins are spread and scattered across the DNA and even across different chromosomes. To piece these fragments back together to form one readable mRNA needs another process and another enzyme.

When a protein is needed and the DNA transcribed into RNA, all the useful coding introns and the non-coding exons are also copied. This is known as the primary RNA transcript. Before it goes out of the nucleus to find a ribosome to make a protein, it needs some modification. The RNA needs to be sliced into the usable exons and non-usable introns and then the useful segments are joined back together again. This is called RNA 'splicing'. This is similar to photocopying an entire chapter of a book but only needing a handful of pages or diagrams. The useful pages are taped together and then laminated as mRNA. The waste is left behind to be recycled. The RNA then receives its 'hat and coat' protection before being 'posted' through the nuclear pores to leave the nucleus and enter into the cytoplasm to find a ribosome.

As we have seen, there is quite a difference between the methods employed to store the information between prokaryotes and eukaryotes. As a result, there is a marked difference in the speed of producing RNA and translating it into a protein. In a prokaryote, the mRNA is made from a continuous piece of coding DNA and, as this is done in the same place where the ribosomes are found, both transcription and translation can be done simultaneously. In eukaryotes, the process is slower. The RNA is made, but next it needs to be cut up (to remove the introns), and the useful bits (the exons) reorganised and stuck back together. It then needs to have its protective cap and polyA tail added before leaving the nucleus. There is a physical separation between the site of RNA production and the ribosomes to make the protein. The advantage for the eukaryotic cell is that the mRNA will survive longer due to its cap and tail; this reduces the speed with which it will be digested by enzymes that break down RNA into individual nucleotides called nucleases.

Nucleases can be useful. If a piece of mRNA were made that had been incorrectly produced or damaged in transit, then a nuclease is needed to break it back down and start again, in the same way a Lego kit can be broken apart and reassembled according to the original instructions. Stopping mRNA being translated into erroneous proteins saves wasting energy and resources. As a result, there is an ongoing process of creating and then destroying mRNA by mechanisms within the cell. There are other enzymes in the cytoplasm that help with housekeeping; they remove superfluous molecules such as proteases, which break down and remove malformed proteins.

Prokaryotes have another advantage, as they are able to cluster several useful genes next to each other on their DNA. Activation of the gene allows several proteins to be made simultaneously. When several genes are kept together they are known as polycistronic genes (poly = 'many'; cistron = 'segment of DNA that specifies a single protein'). We saw the activation of the lac operon in the previous chapter, and we will consider the energy store of ATP in the next chapter. Several different proteins are needed to generate ATP. Just as the sugar, flour and cocoa are kept together in the kitchen for convenience when making a cake, keeping all the 'recipes' or genes together for these proteins makes sense. Coordinating production of the RNA and proteins quickly and easily gives prokaryotes a reproductive advantage.

Problems with missing exons

Some diseases have whole sections of their DNA missing. In muscular dystrophy, there is a problem with a part of the gene on the X chromosome. There should be a code for a protein called dystrophin, which has 79 exons. When these useful pieces of DNA are linked together they ultimately form one readable piece of mRNA waiting translation into a protein. Dystrophin is a protein mainly found in muscle cells and is thought to be a protector of the muscle fibres. When this gene is abnormal, it causes the disease because there is an exon missing. In Becker muscular dystrophy, the rest of the exons can still be knitted together, but the protein produced is abnormal. If the exons cannot be joined together, the protein cannot be produced and is missing, such as in Duchenne muscular dystrophy. The lack of dystrophin leads to progressive accumulative damage to the muscle fibres and ultimately the whole muscle. As a result, the muscle cells weaken and sufferers die because of muscle weakness, problems moving about and heart problems. New medical advances and breakthroughs include 'exon skipping' and CRISPR Cas9 technology. Exon skipping allows the mechanisms of RNA production to 'jump' over the missing exon and join the two pieces of mRNA together and so be translated into a functioning protein. There is more on CRISPR Cas9 technology in Chapter 14.

Summary

Genes are used to create proteins and we have read about one famous example using *E. coli* and lactose. Problems can arise if genes are inappropriately expressed or if mistakes are made. We have seen the different types of mutation which can arise.

Chapter 4 Mitochondria, ATP and endosymbiosis

Over the long term, symbiosis is more useful than parasitism. More fun, too. Ask any mitochondria.

– Larry Wall

What is ATP?

So far we have seen several processes, such as the manufacture of proteins, which need energy. Cells, rather than having a charged battery as an energy store, use a molecule called adenosine triphosphate (ATP).

ATP is made up from adenine (which we met earlier as one of the four bases of DNA) attached to a ribose sugar that can be joined to up to three phosphate atoms. Attaching phosphate atoms to the adenosine stores a lot of energy. One phosphate atom stores a little energy, but having two gives more, while the third phosphate atom stores the most energy. Energy is released and made available to the cell by letting go of the third phosphate atom. Having ATP in a cell is like storing electricity in a battery – the energy is there when it is needed.

The name of the molecule changes as ribose and more phosphate atoms are added to the original adenine. Adenine and ribose make adenosine. Adding a single phosphate atom creates adenosine monophosphate or AMP ('mono' = single); a second generates adenosine diphosphate or ADP ('di' = two); and the third phosphate atom makes the ultimate energy store of adenosine triphosphate, aka ATP ('tri' = three). This is demonstrated in Fig. 1.

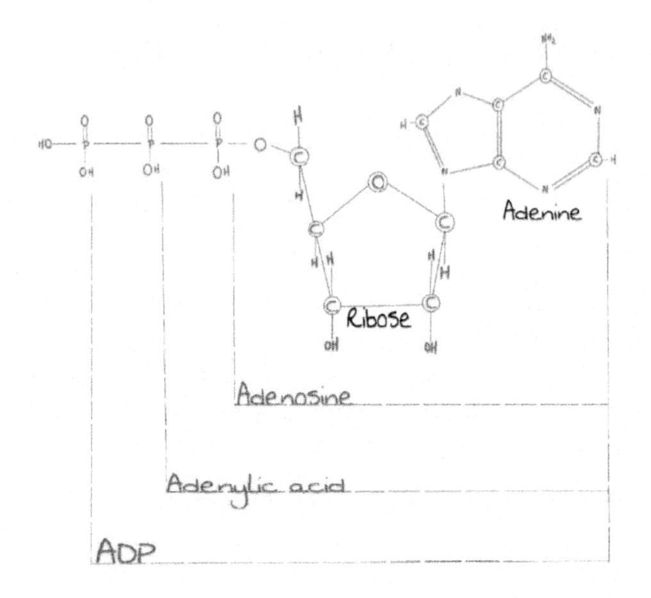

Fig. 1 Combining adenine with ribose and phosphate atoms to form ATP

Why is ATP important?

Over the millennia, humankind has created a small, easy-to-carry, universally accepted series of metal coins and paper banknotes. Money is used to place a value on the time the plumber spends repairing your boiler, or the pizzas we have delivered, or the groceries we shop for. Using multiple ways to value time spent or objects bought or a system of bartering would lead to great difficulties.

ATP is the universal currency of energy for all cells on Earth. ATP can be seen to have similar properties to money. It supplies a good amount of energy; it releases the energy steadily over time rather than in high bursts; and, importantly, ATP isn't toxic in high concentrations.

The phosphate atom can be added or removed as energy is available or as it is needed respectively. The energy can be made available to a variety of different proteins and different processes within the cell. Starting as ATP, when a phosphate atom is lost it changes to ADP. More energy can be wrung out of the ADP molecule by removing another phosphate atom to finally become AMP. ATP isn't the only energy carrier - there are specialised carriers in some cells - but it is by far the most common.

The powerhouse of the cell

ATP is made by special organelles within eukaryotic cells called mitochondria (singular = mitochondrion). A simple diagram of a mitochondrion is shown in Fig. 2.

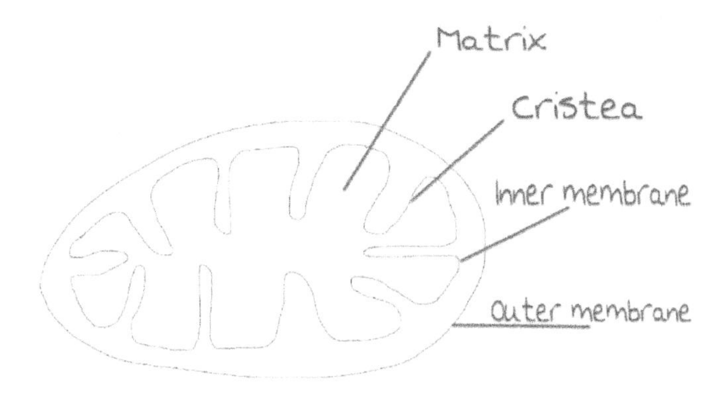

Fig. 2 Simple mitochondrion

The cristae are the folds in the inner membrane that increase the surface area. When sugars such as glucose are in the mitochondria and combined with oxygen, there is a release of energy. This energy is stored as ATP for the cell to use at a later date. Carbon dioxide is released as a waste product. In plant cells, there is an additional organelle called a chloroplast, which uses energy from the sun and stores energy as sugar molecules. Later in this chapter we will see how the mitochondria and chloroplasts arrived in the cell.

The mitochondrion is not an empty spherical space: it has a highly folded internal phospholipid membrane. This folding increases the surface area and tucks it into a small space. The space between the outer and inner membranes is the intermembrane space. This is where the magic of energy generation and storage happens. The internal membrane crams as many energy-generating units into itself as possible, as seen in Fig. 3. Each mitochondrion has some DNA and ribosomes of its own so it can make some useful proteins. The genes of a mitochondrion are considerably smaller than those stored in the cell nucleus. The human mitochondrial genome has only 37 genes, whereas the nuclear genome can contain 35,000 genes.

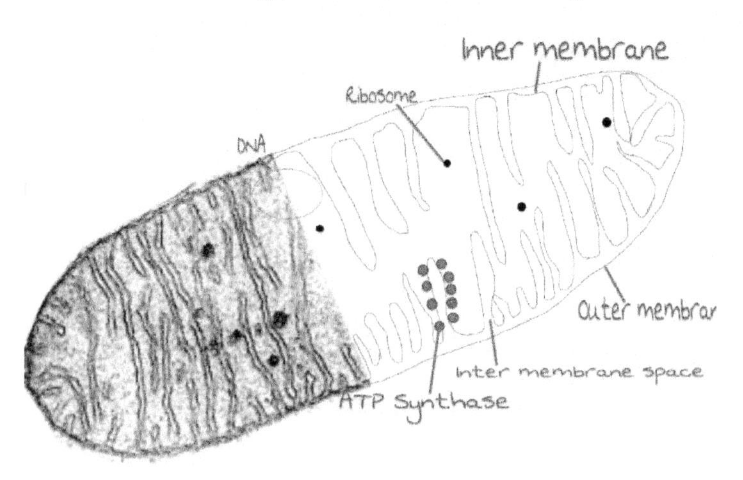

Fig. 3 More complex mitochondrion

A cell may have many thousand mitochondria, each with the machinery to make thousands of ATP molecules. A human liver cell has between 1,000 and 2,000 mitochondria and they take up to about 20% of the cell's volume. In a muscle cell, it may be as high as 40% of the cell. Being able to produce energy gives these cells a huge advantage.

Generating energy

To store energy as ATP requires a special generator. The energy-generating units are called ATP synthase and are made of a bunch of proteins working together.

We have seen that sugars are made from carbon, hydrogen and oxygen atoms. Hydrogen can be broken down into a proton (also called H^+) and an electron (known as e^-). When combined, one proton added to one electron would form a single hydrogen atom. Equally, a single hydrogen atom can be broken down into the two constituent parts.

ATP synthase works by pushing protons through its turbine. Just as a child on a helter-skelter spirals round and round as they whizz back to earth, a proton passing through the ATP synthase makes *it* whizz round. As ATP synthase rotates, an ADP molecule and a phosphate atom are forced together to create ATP with the addition of the spinning energy. In effect, the spinning energy is being stored in the bond created between the phosphate and the ADP. The ATP can be transported elsewhere in the cell and the energy released at a place and time when needed.

The 'turbine' straddles the inner membrane of the mitochondrion. To have protons whizz through, a sufficient number of protons need to be on the correct side of the membrane. A large number are pushed from the inside of the mitochondrion across the inner membrane and into the intermembrane space.

Impressively, so many protons are concentrated that the central rotor of each ATP synthase turns about 150 times per second! Every day each of us produces an amount of ATP equivalent to our own body weight to provide the energy needed to sustain our life.

ATP is used in all organisms, but the types of ATP synthase vary between organisms. For example, *E. coli* has the simplest form of ATP synthase.

The ATP generated is able to leave the mitochondria for use by other parts of the cell. Carbon dioxide is also produced as a waste product. As ATP and CO_2 leave the cell, so ADP, phosphate and oxygen enter as raw ingredients for the next cycle of ATP production.

More details and getting complicated

The equation of respiration shows that we use oxygen and glucose and produce carbon dioxide as a waste product.

Glucose ($C_6H_{12}O_6$) + $6O_2$ -> $6CO_2$ + $6H_2O$ + energy (stored as ATP)

This glucose-to-energy process is much more complex than written here and uses many different molecules. It can be called the Krebs cycle, citric acid cycle or the tricarboxylic (TCA) cycle. For now, we need to know that the end product is a molecule that is a hydrogen carrier – although, technically, it is a proton carrier as H^+ is a hydrogen atom that has lost its electron. The proton carrier is called nicotinamide adenine dinucleotide (NAD). When combined with a proton, NADH is produced. NADH then 'donates' electrons to generate

ATP. These compounds - NAD and NADH - cycle back and forth, gaining and losing protons as in the equation:

$$NAD^{+} + H^{+} + 2e^{-} \longleftrightarrow NADH$$

In actuality, when NAD holds on to the proton it needs two electrons. This effectively produces H^- consisting of one (positive) proton (H^+) combined with two negative electrons ($2e^-$), leading to an overall charge of H^-. When the proton is freed from NADH, two electrons enter the ATP production chain. The NAD returns to the Krebs cycle to gain another proton and two electrons. NADH effectively is a proton and an electron carrier simultaneously. If there are no electrons from NADH, then there is no proton movement across the membrane and no ATP is generated.

ATP generation

On the inner membrane of the mitochondrion sits three protein complexes, which together are called a 'respiratory enzyme complex' or an 'electron transport chain'. The electrons from the NADH flow through the three complexes. As the electrons flow, the protons are transported across the inner membrane and into the intermembrane space, and the concentration of protons in the intermembrane space increases.

We saw previously that atoms and electrons 'want' to be in the lowest possible energy state. Imagine the electron as a skier at the bottom of a mountain that they want to ski down. If they are rich, they can jump into a helicopter and be flown to the top of the mountain. In doing so, they gain (potential) energy. The electron gains energy from the $NADH \rightarrow NAD^{+} + H^{+} + 2e^{-}$ reaction. The skier starts in a high-energy state and starts to release their energy as they ski down the mountain. The energy the electron has is released as it passes each of the respiratory enzyme complexes. These protein complexes use the energy released to transport protons across into the intermembrane space of the mitochondrion.

The electron travels down through the three electron transport chain enzymes but can't do it unaccompanied. Imagine letting an eight-year-old ski down a mountain unsupervised! The helping hand is seen in Fig. 4. The first minder to transport the electron safely is known as ubiquinone (or 'Q', but sadly nothing to do with James Bond). From here the electron is handed on to the next bodyguard - cytochrome c. This escorts it to the bottom of the ski slope, where it is handed on to oxygen and used to form a water molecule. By this time, it is exhausted and has given up all its energy on the way down.

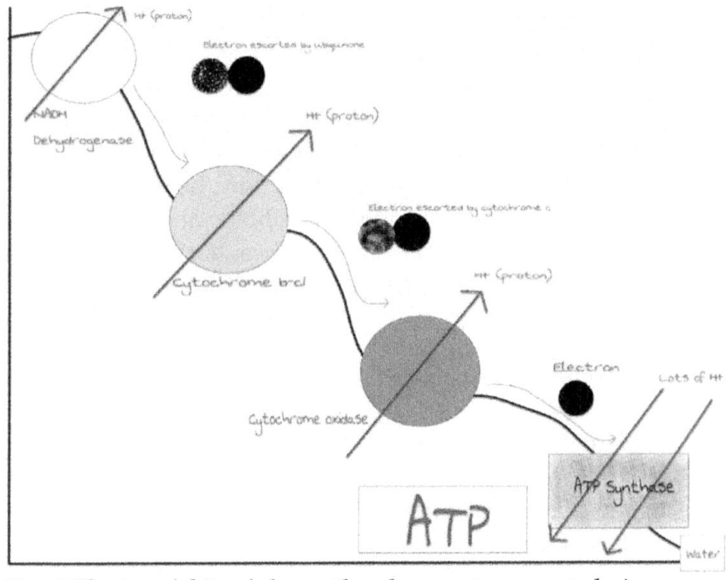

Fig. 4 Electron 'skiing' down the electron transport chain

The electron transport chain proteins

NADH dehydrogenase is the first protein complex of the electron transport chain. It consists of 22 proteins and frees the proton from NADH; in doing so, it also releases two electrons. These electrons are carried by ubiquinone until it reaches the second complex. NADH dehydrogenase transports four protons to the intermembrane space with the energy it gained from the electron.

The second complex is cytochrome b-c1 complex (aka ubiquinol-cytochrome c reductase) and is made of 400 amino acids. This too is able to transport protons into the intermembrane space from inside of the mitochondrion as the electron passes. It passes the electron on to cytochrome c to reach the third complex.

The final complex is cytochrome oxidase complex. This transports a further four protons to the intermembrane space. At the end of the electron transport chain a kindly oxygen atom accepts the electron in the low-energy state and uses it to form water. Oxygen is therefore known as a 'terminal electron acceptor'.

Poisons and ATP

Importantly, if you are ever poisoned by cyanide, it is worth knowing that it blocks the activity of the electron transport chain - specifically, the final protein (cytochrome oxidase) complex. In effect, cyanide turns off one of the proton-pumping stations. If no protons flow, then no more energy can be formed by the cell.

Other poisons also block the electron transport chain: rotenone blocks NADH dehydrogenase; and antimycin A blocks cytochrome b-c. Interestingly, rotenone is made naturally by some plants and is used both as an insecticide and a pesticide. 'Proton uncouplers' such as dinitrophenol allow protons back into the cell without having to go through the ATP synthase. Dinitrophenol was used historically in some diet pills, but is incredibly dangerous. It kills by leaving no outlet for all the energy produced by the mitochondria, so the person who consumes it can end up 'cooking' themselves to death. ATP cannot be produced if the proton channel of ATP synthase is blocked by a naturally occurring antibiotic called oligomycin.

ATP production

Assuming we have not been poisoned, lots of protons have been transported into the intermembrane space, leaving far fewer inside the mitochondrial matrix. This means there is a gradient across the membrane. The gradient consists of both an electrical gradient (due to the positive and negative charges) and a pH gradient (due to the movement of protons). The electrical gradient is more important; it is the driving force in the generation of ATP and is known as the proton motive force.

The collection of proteins of ATP synthase sits across the inner membrane of the mitochondrion and acts as a generator of ATP as the protons whizz past. As the membrane is impermeable to protons, the only way they can return to the inside of the mitochondrion is via the ATP synthase complex. ATP is generated as the protons pass through.

$$\text{ADP} + \text{phosphate} + \text{energy} \rightarrow \text{ATP}$$

The rotor of ATP synthase can be in one of three positions. The first position is 'open' – where the site is empty. This rotates to 'loose', which is where the ADP and phosphate enter. This is followed by 'tight', where the ADP and phosphate are forced together to form ATP and released. This returns the site to 'open' again.

This is like a child making a snowball. Initially, their gloves are empty or 'open' to receive. They take a pile of snow in both hands and hold it 'loosely'. The snow is forced together 'tightly' and forms the finished article - a snowball - which is released!

In plant cells, chloroplasts are used to generate ATP. Chloroplasts use light to split water into hydrogen and oxygen. The electron and proton from the hydrogen pass into a system of proteins not dissimilar to the respiratory enzyme complex of the mitochondrion. The differences are that oxygen is the waste product rather than carbon dioxide and that the original energy comes from light rather than glucose.

Endosymbiosis theory

Mitochondria are very efficient producers of energy, but how did the cells end up with these generators inside them? To understand the current theory, we need to go back in time to revisit some of the earliest cells.

Initially, some prokaryotes - bacteria and archaea - produced energy without oxygen, as the Earth was an oxygen-free zone. They could generate ATP via light (phototrophs) or chemicals (chemolithotrophs). This is called anaerobic respiration (or '*without* oxygen'). Initially, chemolithotrophs used a variety of chemicals such as sulphur, iron and methane to make ATP. Subsequently, about 2.7 billion years ago, cyanobacteria evolved. These were phototrophs and used sunlight to generate ATP via a protein cascade similar to that of the respiratory protein complex used by current-day eukaryotic cells. These cells produced oxygen as a waste product from splitting water. It has been argued that these were the most important cells to have evolved on the planet. All photosynthetic cells such as plants and algae share a common ancestor with these cyanobacteria. However, for a billion years or so, despite oxygen being produced, the levels in the atmosphere remained stubbornly between 1% and 2%.

Why was this? Oxygen levels could only increase when production exceeded consumption. First, the oxygen became dissolved in the water of the oceans. Once this was saturated, the oxygen left the oceans, but there were many other elements on the land that bound to and reacted with it. It took until about 850 million years ago that oxygen accumulated in the atmosphere and levels rose enough for more complex creatures and animals to appear.

The climbing oxygen levels allowed a sudden burst of evolutionary activity. Bacteria that were able to make use of the oxygen did so by means of aerobic respiration (or *with* oxygen). Using oxygen is a much more effective way of generating ATP. As a result of excess energy production in cells, new opportunities arose. The proliferation of life at this time is known as the 'Cambrian explosion'. Currently, there is 21% oxygen in the air. However, oxygen levels reached a staggering 35% about 360-300 million years ago during the Carboniferous period. In part this is thought to be due to the huge expansion in the amount and variety of plant life.

Some organisms couldn't cope with the increase in oxygen levels so either became extinct or found low-oxygen niches such as deep oceans to survive in.

There were anaerobic bacteria and archaea - some using chemicals and others using light to generate energy. Based on the analysis of genomes, there is increasing evidence that the eukaryote cell arose through the fusion of a bacterium and an archaeon about 1.5 billion years ago. The trapped bacteria eventually evolved to become mitochondria. This is called the endosymbiosis or symbiogenesis theory ('sym' = together; 'bio' = living; 'genesis' = birth). Later, on a separate occasion, a photosynthesising (phototrophic) bacterium entered a eukaryotic cell and evolved into a chloroplast.

In 1970, biologist Professor Lynn Margulis suggested the endosymbiosis theory as a mechanism for the origin of mitochondria and chloroplasts from free-living bacteria as they became engulfed by a single-celled archaeon. The energy generated from this newcomer into the cell enabled eukaryotes to evolve and become larger and more complex than the prokaryotic cells.

Humans consume food to gain energy. There is more on food webs in Chapter 10, but for now consider a plant having used energy to create sugars and carbohydrates. For example, the wheat plant takes sunlight and stores this by creating complex carbohydrates called starch. When humans consume wheat, the carbohydrates are broken down into simple sugars, which are subsequently used to generate ATP via ATP synthase, as we have seen.

An amoeba is the simplest eukaryote. It consists of a single cell and it can engulf a prokaryote as a food source and digest it. In effect, the small prokaryote becomes trapped in the large eukaryote cell in a 'bubble' or vacuole. The host cell pours digestive enzymes such as proteases inside the vacuole to destroy the cell wall and fats, proteins and carbohydrates. The host benefits as it reuses and recycles these building blocks for its own energy needs.

Endosymbiosis theory describes a time when one cell engulfed another cell *but didn't destroy it*. The cell survived inside another cell. The internal cell gradually evolved into what is now called a mitochondrion.

Evidence for endosymbiosis theory

There are several things that would have happened when one cell entered another and became trapped. First, the trapped smaller prokaryote would have been enveloped in the outer membrane of the host cell. This would give the appearance of the cell being inside another membrane - in effect, creating a prokaryote with two cell membranes. This is what is seen with the mitochondria: there is an inner membrane and an outer membrane. This is akin to having a chocolate fountain and a marshmallow. Imagine the flowing sheet of chocolate is the eukaryote membrane. The smaller bacterium – the skewered marshmallow - already has an outer membrane, or marshmallow 'skin'. When pushed through the fountain, the marshmallow is coated with liquid chocolate – adding a second membrane. The outer membrane of the bacterium becomes the inner membrane of the mitochondrion. Between these two membranes the intermembrane space has been created. As we have seen, ATP synthase is located across the inner membrane and protons now accumulate in the intermembrane space.

So, there are two membranes, but there must be other reasons to think an archaeon engulfed a bacterium. Mitochondria also contain DNA and ribosomes. The DNA is used to create some of the proteins that are needed in the electron transport chain. The DNA is arranged in a circle – just as in bacteria – and the DNA is not spooled for safe storage - just as in bacteria. Striking similarities between the genes in mitochondria and in bacteria have been found. The mitochondrial genes are found to be degenerate versions of the corresponding bacterial genes.

The ribosomes of mitochondria are more similar to those in bacteria than in eukaryotes. The ribosomes in our eukaryotic cells, and in other animals, plants and fungi, are much larger than their bacterial equivalents. In eukaryotes, they are termed 80S ribosomes. The mitochondria have smaller 70S ribosomes, which are similar to bacterial ribosomes.

Bacteria replicate by a process called binary fission. This means the cell splits into two identical daughter cells. Mitochondria also replicate by means of binary fission.

Mitochondrial genes

Ribosomes are made of proteins and so are coded for by ribosomal RNA (rRNA). Ribosomes are made of two subunits. The smaller subunit is coded for by the 16S rRNA gene. This is a very important gene because it is needed in all cells to create ribosome. Without it, no ribosome can be made, so no protein, which means the cell can't survive. Analysing this gene and comparing the small changes found can help show which organisms are related to each other genetically. Analysis of the 16S rRNA gene from mitochondria shows it to be very similar to bacterial 16S rRNA. (See Chapter 7 for more on the 16S rRNA gene.)

These observations led to the hypothesis that mitochondria (and chloroplasts in plant cells) were actually bacteria that were caught inside cells early in the evolution of eukaryotic cells.

Mitochondrial DNA is very useful for showing evolutionary history, as it is only passed from a mother to her offspring. Changes in the sequence of mitochondrial DNA are much slower to accumulate and can be used to show evidence of human migration patterns.

The genes found in mitochondria are interesting. The whole gene length is 16,589 base pairs. It codes for some proteins as expected, but only 13 of them. This means that there are genes for *some* of the proteins needed in the respiratory transport chain *but not all of them, some* proteins of ATP synthase *but not all of them* and *some* proteins of the ribosome - specifically 16S rRNA and 12S rRNA - *but not all of them*. The other proteins needed to complete the electron transport chain and ATP synthase are made by the host itself - i.e. there are genes in the host's nucleus that code for mitochondrial proteins. The host produces RNA and subsequently proteins via its own ribosomes for the mitochondria to use.

One hypothesis is that the mitochondrion retains genes that make at least one essential protein for each part of the respiratory enzyme chain. This way it retains local control and can regulate its own energy production. Another hypothesis is that the mitochondrial DNA codes for proteins are hydrophobic. This means that the proteins could get 'stuck' in transit to the mitochondrion if they weren't made by the mitochondrion itself.

Why would a bacterium 'want' to live inside a host cell?
A bacterium just 'wants' to survive long enough to reproduce and pass on its genes to its daughters - the 'selfish gene' coined by Richard Dawkins. Living inside a host cell dramatically changes the environment. Typically, the internal environment of a host cell is more or less constant and there are less likely to be extreme

changes, for example, in nutrient level or pH. This has an impact on the selection pressures acting on the mitochondria. If the environment is stable, there are genes that are no longer needed or expressed. These become redundant can be lost to the DNA of the host cell. Single celled prokaryotes such as bacteria tend to jettison genes which are not used but are able to regain these lost genes, as we will see in chapter 7, during horizontal gene transfer.

A cell requires energy to replicate the DNA. If it were to replicate genes that were not being expressed, this process would waste its hard-earned ATP. As the DNA shrinks and unnecessary genes are deleted, there is a noticeable impact on the speed of mitochondrial replication; i.e. with less DNA, replication is faster. If some cells are able to replicate faster than others then, over several generations, they would swamp their slower-reproducing neighbours. They would use the food, nutrients and other resources in the surrounding area. As a result, only those mitochondria with smaller, leaner, more streamlined DNA will remain inside the cell. The more genes that are transferred out of the mitochondria, the faster the mitochondria can replicate.

Some proteins are still essential to the functioning of the mitochondrion. If the mitochondrion can incorporate its own DNA into the host's DNA, it will still benefit from the production of the vital proteins but without having to carry the genes itself. We can see that this is a mutually beneficial relationship. The bacterium-cum-mitochondrion has a safe and stable environment in which to live, and the host cell has access to a huge supply of energy in the form of ATP.

So why have any genes at all in a mitochondrion?
Unfortunately, if there are no genes at all, the mitochondrion fails to reproduce and thrive and therefore can't respire. This means the cell ends up needing to generate ATP from a different source such as anaerobic glycolysis, which is less efficient.

Some proteins may be difficult to get into the mitochondrion if they were made elsewhere. If the mitochondrion lacks a specific protein and is waiting on the host DNA to be transcribed into RNA, then translated to the protein before delivery, it may die due to the time delay. Also, how could the protein know which mitochondria to travel to? There is no satnav for a protein to find its way and, as there can be thousands of mitochondria in a liver or muscle cell, knowing which cell was lacking the required protein would be an impossible task. The most efficient way for host and mitochondrial cell to survive appears to be by sharing genes. In some organisms, the mitochondria have retained an absolutely bare minimum of DNA. There are only five genes in the mitochondria of the malaria-causing parasite *Plasmodium,* which we will meet in Chapter 8.

Obligate intracellular bacteria
Which bacteria is thought be the guilty party that became a mitochondrion?

Rickettsia prowazekii (R. prowazekii) is a bacterium that causes the disease typhus. It was named after Howard Ricketts, who, having discovered the bacteria, later died from the disease. Typhus is a severe illness found in developing countries where there is poor sanitation, poverty and close human contact. It used to be known as 'gaol fever' and was endemic in overcrowded prisons. It causes a fever, rash, joint pains, abdominal pains and diarrhoea, and a cough. It can be fatal if untreated, but modern-day antibiotics are able to cure the disease. *R. prowazekii* is transmitted by the faeces of lice, ticks and fleas. Humans are the main carrier or reservoir of the bacteria. The only other animal known to have *R. prowazekii* is the flying squirrel in America.

Importantly, when *R. prowazekii* attacks, it lives inside host cells. It is called an obligate intracellular parasitic bacterium. This means it is obliged to live inside another cell and cannot survive for long outside a host. It is known as a α-proteobacteria. The genome of *R. prowazekii* is just over one million base pairs long and codes for 834 proteins. The gene codes for NADH dehydrogenase, cytochrome reductase and cytochrome oxidase, as well as those for ATP synthase. These are the proteins that are involved in ATP generation in modern-day mitochondria. As a result of the analysis of the genetic code sequence, an ancestor of *R. prowazekii* is thought to be the bacterium that became the first mitochondrion.

Other intracellular bacteria have been identified that live hand in glove with their host cells. *Rhizobia* are bacteria that live in the soil; their role is to fix nitrogen, as we will see in Chapter 10. They infect the root cells of some plants. As a result, they supply the plant with nitrogen, which is vital for growth, while the plant supplies them with nutrients in the form of carbon and energy. These bacteria are also α-proteobacteria that have formed a symbiotic relationship - that is beneficial to both parties.

Buchnera aphidicola (*B. aphidicola*) is another α-proteobacterium that invades cells, this time of the aphid. They have become so dependent upon each other that neither the bacteria nor the aphid can survive without the other. The aphids live on plants that have little protein. Humans have nine amino acids that we can't make ourselves; we need to get them from our diet. These are known as essential amino acids. Aphids have ten essential amino acids, and these are supplied by *B. aphidicola.* If the bacteria are killed, the aphids fail to make the proteins that are needed to grow or reproduce. What does the *B. aphidicola* get in return? The aphid supplies energy, carbon and nitrogen to the bacteria. The bacteria take up the nutrients and are enabled to grow and reproduce. The evolutionary relationship between these two organisms has been shown by examining the 16S rRNA gene. In a similar way to the transfer of genes from the mitochondria to the host, there have been genes swapped from the *B. aphidicola* DNA to the host nucleus.

Summary

About 1.5 billion years ago, the ancestral protoeukaryote acquired a bacterium. This gradually evolved into a mitochondrion and is found in all eukaryotes. Mitochondria have their own DNA and replicate by fission. Mitochondria are passed from parent cell to daughter cell during cell division. Over time, many genes originally part of mitochondrial DNA have moved to the nucleus of the cell.

Now, mitochondria live and reproduce inside cells, focusing on energy production and relying on the surrounding cell for most of their other needs.

Chapter 5 - Evolution, trees and genetic inheritance

In microbiology the roles of mutation and selection in evolution are coming to be better understood through the use of bacterial cultures of mutant strains.

– *Edward Tatum*

What is evolution and how did it occur?

Single-celled microorganisms were the first living organisms on Earth. Subsequently, they evolved into the vast diversity of life we see around us.

Despite the diversity of life, living things are essentially the same on the inside. The cell is the fundamental unit of life, and we have seen how, on the basis of cell structure, organisms are classified into two groups: prokaryotes and eukaryotes.

As far back as the ancient Greeks, people have tried to understand where life came from. Prior to 1859, the main view of the world was creationism - that life had developed through the guiding hand of a divine being. Some believed that God had created life perfectly, and that the species were eternal and did not need to change or evolve. This also meant that there would be no need for extinctions or loss of species.

In 1859, Charles Darwin proposed the theory of natural selection in his work *On the Origin of Species*. In scientific terms, a 'theory' is defined as 'a well-substantiated explanation of some aspect of the natural world that is acquired through the scientific method and repeatedly tested and confirmed through observation and experimentation' rather than a hunch or an idea.

Scientifically, hunches and ideas are called hypotheses, which are tested repeatedly to become a theory. Scientific 'theory' should really be regarded as fact. In the 150 years since the theory of natural selection was first expressed, there have been many advances in the scientific community that have helped to confirm Darwin's findings, such as genetic technology.

Darwin's postulates

Darwin's basic principles or postulates were that:

- The history of the Earth is long.
- Individuals are variable.
- Some variations are passed down.
- More offspring are produced than can survive.
- Survival and reproduction are not random.

We need to understand these postulates in a little more detail.

The Earth is old

The Earth is about 4.5 billion years old. Cellular life has been present for at least 3.4 billion years and perhaps for as many as 4.1 billion years. Fig. 1 represents the entire history of the Earth if it were compressed into a single 24-hour period. On this scale, ancient upright humans would have appeared at 23.59, or one minute before midnight. In real time, this represents about 200,000 years. Modern humans arrived at just four seconds to midnight as our hands evolved to grasp, our shoulder joints rotated, our brain size increased and the knee supported upright walking. In that time, we changed from being cave dwellers to becoming farmers about 10,000 years ago. Since then, the development of civilisations has accelerated to shape the world we know today.

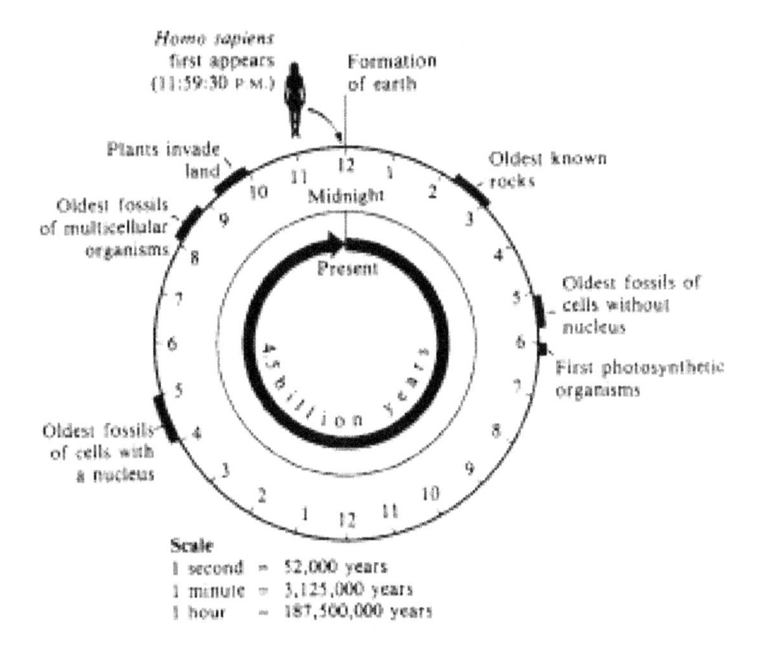

Fig. 1 The history of the Earth in 24 hours (Credit - tatfoundation.org/24hour.gif)

On the 24-hour clock, life itself began at about 8am with a common cellular ancestor. This cell, from which all modern-day life descended, was the Last Universal Common Ancestor (LUCA). This was a simple prokaryote, which we will hear more about later in the chapter. These cells did not use oxygen but gained energy first from chemicals and later on from light.

At about 10am, bacteria evolved the ability to split water into oxygen and hydrogen gases. As a result, oxygen levels rose rapidly. 'Rapidly' still took a billion years!

The complex eukaryotic cell descended from the prokaryotic cell just once in 4.5 billion years. The endosymbiosis event that we saw in the last chapter occurred approximately 1.5 billion years ago.

By 6.40pm, algae had appeared and were among the first plants. Between 9 and 10pm, trilobites, land plants, fish and insects all arrived. Most of this life was wiped out in the largest extinction event known, at around 10.40pm. By 10.47pm, the dinosaurs were walking the Earth, and, after just nine more minutes, mammals arrived. Dinosaurs became extinct at about 11.40pm and left the gap that mammals, apes and then humans came to fill as the dominant species.

Individuals are variable

There is variability within a population. Some bacteria - such as *E. coli* - can divide six times a day in optimum conditions. As we have seen, mistakes in the copying of the genetic code can cause changes to the cell which introduces variability. Each of these changes on its own is tiny. One base pair here or there may or may not matter much, as we have seen. However, when the changes are combined together and accumulate over many generations, the effects can be profound. Some of the variations will be beneficial and some detrimental to the survival of that cell.

Bacterial cells divide into two daughter cells by binary fission. By examining the sequences of DNA, changes or mutations can be spotted. Fig. 2 shows the changes to an exceedingly short segment of DNA over three generations. The original bacterium divides and a single nucleotide suffers a substitution mutation - the changing nucleotide base is underlined. One of the daughter cells can be seen to have had a change prior to the second division. If after the second generation there is a further base pair change in one of the daughters but not in the other three, then by the third generation of cells a more varied picture appears. There have now been a total of ten base pairs changed across the eight daughter cells.

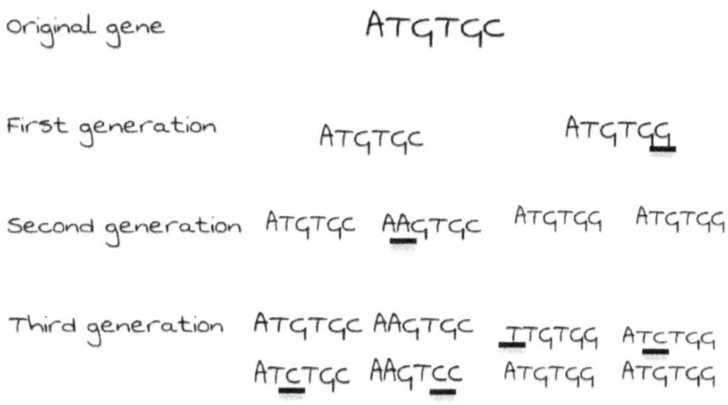

Original gene ATGTGC

First generation ATGTGC ATGTGG

Second generation ATGTGC AAGTGC ATGTGG ATGTGG

Third generation ATGTGC AAGTGC TTGTGG ATCTGG
ATCTGC AAGTCC ATGTGG ATGTGG

Fig. 2 Mutations in genes

If a modern-day cell had the sequence AAGTCC in the DNA we can work out its ancestral line. By comparing this to the four second-generation organisms, we can determine who the parent must have been. Our sequence is found in the lower line of third-generation cells, so its parent must have been a cell with the gene sequence AAGTGC. As mutations accumulate in the cells, their differences become more noticeable.

The organism with the smallest known gene found so far is the bacterium *Mycoplasma genitalium*. It is 580,000 bases long and codes for 482 proteins. DNA replication has a rough error rate of one in a million. So *M. genitalium* would have one error approximately every two generations.

We can use this information to show the variability within populations. Some of the mutations may be lethal, in which case the cell would die and the genes would not be passed on to the next generation.

Some variations are passed down

We can see from the section of DNA above that there is variation in the genes and that this variation is passed to the next generation. The sequence of genes in turn leads to a variation in the proteins that are made by the cell, and these proteins ultimately affect what the cell looks like or how it behaves.

Fig. 3 shows a similar pattern of genetic variation but on a cellular level. Cell (a) divides into two daughter cells (b and c). If there are slight changes during the division process, then one daughter may have a greater or lesser likelihood of surviving. For example, if the cell line (b) has a greater ability to digest a certain food source, and there is more of that food source in the environment, then it will be in a position to reproduce faster and pass its genes on to its daughter cells.

If the cells (d) have lost a gene for making the protein that helps form the cell membrane, then they may end up being smaller in size.

If genes that make an internal structure from improved infolding of the membrane create a better nucleus and improve their internal surface area, these cells would gain an advantage (e).

Some cells may pass on genes that are fatal, so the daughter cell does not survive to reproduce (f). Lethal genes, for example, could be ones that stop the cell membrane occurring, or cause problems with the ability to replicate DNA or repair damaged DNA.

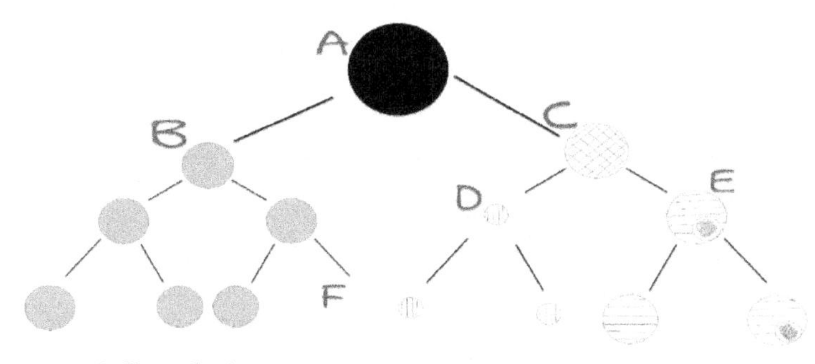

Fig. 3 Cell evolution

For variation to occur it needs to be passed down from parent to offspring. This is known as 'descent with modification'.

More offspring are produced than can survive

The growth of organisms is limited by many factors, such as the availability of nutrients or water or the number of predators. For multicellular organisms such as fish or frogs and trees or flowers, more 'seeds' are produced than will eventually grow into the 'adult' version. Some seeds and offspring will be eaten, others will starve, and plants have a proportion of their seeds that never germinate. In addition, offspring that don't reach reproductive age cannot pass their genes down to the next generation. To ensure survival of any individual species, a great deal more copies of their genes need to be made and distributed.

Survival and reproduction are not random
This is best known as 'survival of the fittest', but could be more accurately written as 'survival of the best adapted'. As there is variation within each population and more offspring are produced than will survive to be able to reproduce themselves, those individuals with the best traits to survive *that environment* will survive and reproduce. Hence, these will be able to pass their genes to their offspring while those with fewer adaptations to survive *in that environment* won't be able to.

An example of this would be the dodo (*Raphus cucullatus*). Dodos were massively overweight pigeon-like birds. They lost their ability to fly over many generations due to the lack of the predators on the island of Mauritius they inhabited. The birds did not need to put time and energy into being able to fly when there was nothing to fly away from. When the Portuguese discovered the island in 1507 they found the dodos easy to capture, as they had no understanding of predators. They had evolved to suit their environment but could not evolve fast enough when it changed.

How did Darwin get the idea of natural selection?
The well-told story of Darwin's trip to the Galapagos Islands will be retold, albeit briefly, here. Darwin joined the survey ship HMS *Beagle* in 1831. The planned two-year journey of discovery turned into a five-year adventure. It was while sailing that Darwin wrote his book *The Voyage of the Beagle*, covering biology, geology and anthropology. Finding fossils and examining rock formations, the ideas he gathered were later refined into his book on evolution.

The finches on the Galapagos Islands in the Pacific Ocean caught Darwin's imagination. There are 15 species of

Galapagos finch, although they are not true finches but belong to the tanager family. The birds vary in size and weight, but it is their beaks for which they became most famous. The sizes and shapes of their beaks became adapted to certain food sources over time. In Fig. 4 we can see some of the finches Darwin identified. The different types of beak are adapted to their function, such as grasping, probing and crushing. The finches that eat seeds live on the ground and have beaks that crush. The insect-eating finches live in the trees and have beaks suited to probing tree bark to find food.

Fig. 4 Darwin's finches

This puts each finch into its 'niche', or place in the environment. This means one species of finch that is an insect eater can live alongside a finch that eats seeds. It becomes problematical if all seed-eating, ground-dwelling finches try to live together unless they eat different size seeds - and this has been found to be the case. Those finches with larger beaks tend to eat larger seeds, while those with smaller beaks tend to eat smaller seeds.

More details
Analysis of the finches' genes has shown that the BMP4 gene is important. This codes for bone proteins and, depending on how it is expressed, results in the variations of beak shape and size.

What did the first tree of life look like?
Darwin recognised that the finches had a common ancestor and drew his first 'tree' on this basis. A copy of his original drawing is shown in Fig. 5. There is a common ancestor, or origin of life, labelled 1. The tree shows the branches but also shows dead ends. Dead ends occur when a species becomes extinct and its genes are not passed on.

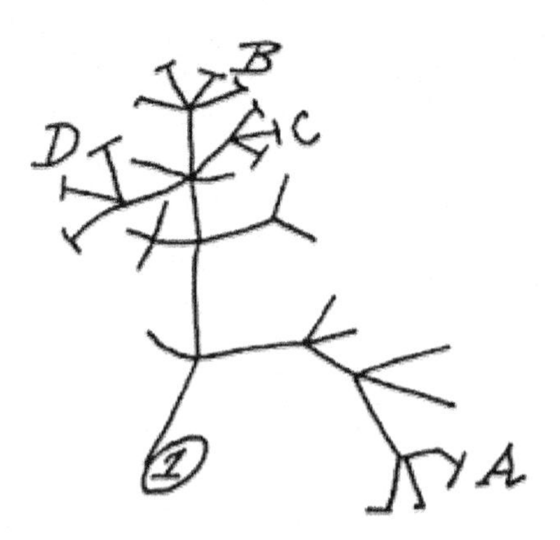

Fig. 5 Darwin's original tree

The Finches – A play to perform

Cast: Mum, dad, son 1 (all with the same beak) and son 2 (thinner beak)
Setting: in the family nest

Son 1: Hi mum, hi dad.
Mum: Hi son.
Dad: Don't we all look the same? Haven't we all got such nice beaks? What beautiful feathers we have. Aren't we lucky in this family to all look the same?

Son 2 enters, looking downbeat with his thinner beak.
Son 2: I feel left out with my beak.
Son 2: Guess who I saw today, brother? I saw that hot chick from the nest in the next tree.
Son 1: You know I fancy her too!
Dad: Now children, don't fight over a girl. You know she is only interested in having your eggs.
Mum: Girl finches like to be treated well, with nice food.
Son 2: Guess what I found today? A beautiful orange nut. Would you like to see it?
All: Yes, please.

The family leave the nest and see the orange nut in the tree.
Son 1: I can get that.
Son 1 tries to get the seed but fails, as his beak **can't** grasp the nut.
Son 2: Watch me!
Son 2 tries to get the seed and succeeds as his beak **can** grasp the nut.
Son 2: I'm going to see the girl next door. It's not the size of your beak, it's the way you use it!
Son 2 flies off.
Son 1: Why does his beak look different from mine?
Mum: I always worried that you would ask about that one day. It's a secret so you can't tell anyone about it. Your brother is ADAPTED!

To put Darwin's finches into a diagram would involve making another tree; Fig. 6 shows how such a tree developed. A common ancestor of the finches is at the base of the tree. This would have laid eggs that then hatched into offspring. As genetic changes occurred, some finches would manage better in certain environments but the different finches are still related to each other.

As more eggs were laid by subsequent generations, the accumulation of changes to the genetic code built up. This slowly changed the shape of the beak to enable it to adapt to the food sources available. These changes were passed from one generation to the next and the genes were inherited by the offspring.

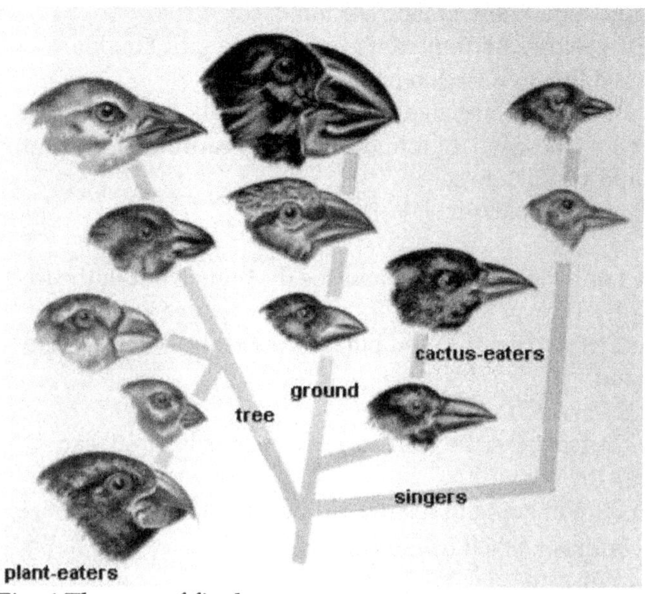

Fig. 6 The tree of finches

Mendel and inheritance
At about the same time as Darwin was publishing his work, the scientist Gregor Mendel was performing experiments on pea plants. Pea plants were chosen for their ability to self-pollinate. Mendel identified seven traits that seemed to be passed on independently of each other from one generation to the next. Between 1854 and 1863, he grew more than 28,000 plants and showed that for each trait there were two copies of each gene. He found that some genes were dominant - that is, if present, they expressed themselves over the other gene in the pair. This is where one copy of the 'book' outranks and overrides the other copy.

Tallness of the pea plant was one trait studied extensively. For example, if a pea plant had a gene for tallness we call it 'T'. As there are two copies of the gene in each plant, we can label them T and T. If small pea plants have the gene 't' then the two copies are t and t.

If these two plants were bred together, both would pass on one gene to their daughters. From the tall plant it must be a T, and from the small plant it must be a t. Therefore, the daughter plants will have gene Tt, as seen in Fig. 8. As the T gene is dominant, then the plant will still be tall. The t gene will be 'silent' or recessive.

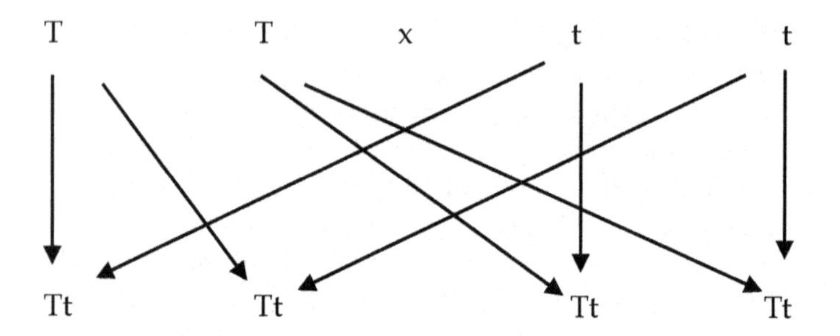

Fig. 8 Cross-pollination of original plants

All the plants will be tall and have gene Tt.

If two of these daughter plants are bred together then each will pass on one gene from Tt, as in Fig. 9.

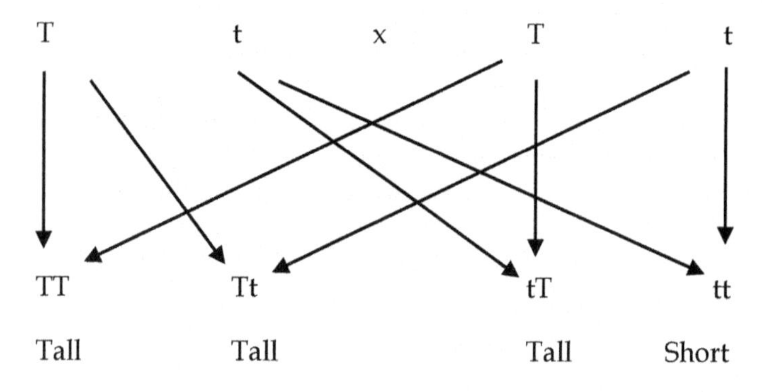

Fig. 9 Cross-pollination of the daughter pea plants

Now, from two tall parents, we have three tall daughters and one short daughter! This is how a genetic trait appears to skip a generation and reappear.

Human eye colour is determined by two genes. The dominant gene is for brown eyes (B) and the recessive gene is for blue eyes (b). Of course, if we have brown eyes we don't know if we have genes BB or Bb as both would show up as brown. On average, if brown-eyed parents had four children then one would have blue eyes. However, some parents would only ever have brown-eyed children if they started with BB genes. In this case, if one parent has brown eyes (BB) and the other is blue eyed (bb), all the children would have brown eyes (Bb). But, if one of these children had a child with another person with brown eyes who had the recessive gene (Bb), there would be a 1 in 4 chance of a grandchild having blue eyes (bb).

This shows that traits or characteristics can be hereditary. However, genes are not the only determinant or influence on the final outcome. For example, for the pea plants to reach their full potential they need nutrients, water and light. Even with the best genes, the environment plays a role. For blue or brown eyes the environment has no influence on final gene expression.

Mendel could have used his findings to develop and selectively breed new strains of pea plant that only had certain characteristics. This is currently applied to other areas of farming, whether cattle, chickens or wheat crops, and is discussed in more detail in Chapter 14.

Naming and classification system

Even before Darwin's book, a system had been created for naming organisms. In 1735, the Swedish naturalist Carl Linnaeus developed the binomial naming system to classify organisms by grouping similar ones together. With this system, every organism is given two names: the genus and the species (a bit like a double-barrelled surname).

Some common organisms listed by the genus followed by the species name include *Homo sapiens* (man), *Mus musculus* (house mouse), *Panthera tigris* (tiger), *Bellis perennis* (common daisy) and *Gorilla gorilla* (gorilla!). The first part or genus is always capitalised, while the second part or species is lower case. Both names are written in italics in scientific circles.

This classification system attempts to arrange species into groups based on their origins and relationships. It is a hierarchical system where smaller groups are placed within larger groups and there is no overlap between groups.

Domains: To start with, there are three domains: archaea, bacteria and eukaryotes. Every living organism fits into one of these three domains.

Kingdom: Within the eukaryotes, there are four kingdoms: plants, animals, fungi and protista. This last group is for all organisms that don't fit into any of the other three kingdoms. All of the species in these kingdoms have a nucleus and organelles in their cells.

Phylum: The next subdivision is phylum. The animal kingdom has about 35 phyla, plants have 12 and fungi 7.

Class: These are only used in the animal kingdom now; the botany classes are rarely used. For example, mammalia is used to signify the class of mammals.

Order: A more specialised group containing, for example, primates or carnivora (carnivorous animals).

Family: An even more specialised group.

Genus: This forms the first part of the binomial name - for example, Homo (man) or Felis (cat).

Species: This forms the second part of the binomial name - for example, *sapiens* as in *Homo sapiens* for human or *catus* in *Felix catus* for the common cat.

Family, order and class vary depending on the scientist who is organising the organisms and there is leeway between them when naming. Fig. 7 shows a classification system for the grey wolf.

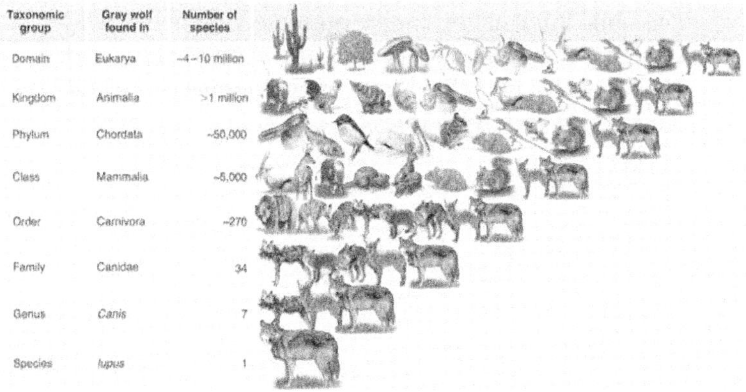

Taxonomic group	Gray wolf found in	Number of species
Domain	Eukarya	~4–10 million
Kingdom	Animalia	>1 million
Phylum	Chordata	~50,000
Class	Mammalia	~5,000
Order	Carnivora	~270
Family	Canidae	34
Genus	*Canis*	7
Species	*lupus*	1

Fig. 7 Linnaeus' classification system for a grey wolf

One way to remember the Linnaeus naming system is with the mnemonic:

Keep	Kingdom
Pond	Phylum
Clean	Class
Or	Order
Frog	Family
Gets	Genus
Sick	Species

Other mnemonics include 'Kids pick candy over fancy green salad' or 'King Phillip come out, for goodness' sake'.

Evolutionary tree

The same information can be presented in another way. Darwin's tree can be further expanded and details added. This is known as a cladogram or evolutionary tree and presents species 'relatedness' on a tree-like structure, as in Fig. 8

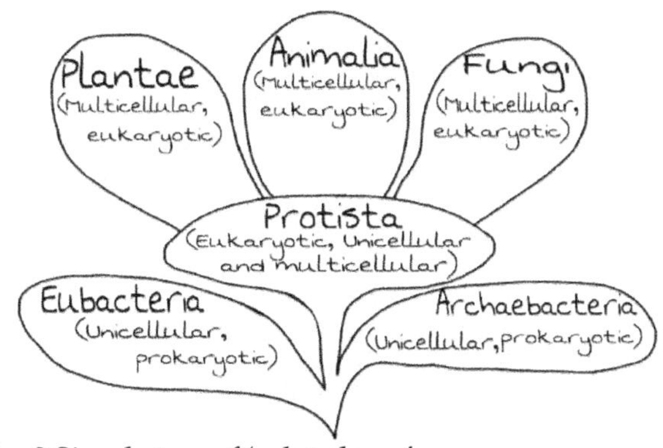

Fig. 8 Simple tree of 'relatedness'

As the trees become more detailed, individual groups or families of organisms can be seen, as in Fig. 9. These groups are formed based on what organisms look like and how they behave. Common features can link species together and can be used to start to trace such different characteristics back in time through previous generations.

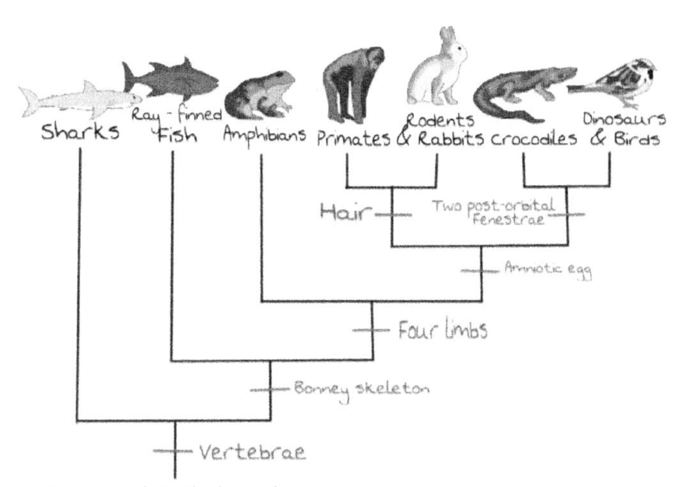

Fig. 9 More detailed evolutionary tree

Extrapolating a tree further and further backwards into the past could take us back 3.8 billion years to our common ancestor, as in Fig. 10. From LUCA, a 'speciation event' is where the branches on the evolutionary tree split in two. The ancestor branches off into two distinctly separate species through mutation of its genetic code.

For example, at point A there was an ancestor that had characteristics now shared with both birds and crocodiles. The common traits include having a vascular system, a nervous system, backbones, jaws, internal organs and the ability to lay eggs. At point A, there was a change to the genetic code that eventually led to birds having feathers rather than scales. This makes birds and crocodiles related in evolutionary terms - at least as cousins.

It is only through the discovery of fossils of extinct species that the process of evolution can be pieced together. At each stage of change there needs to be an inheritable advantage to having the new shapes and structures.

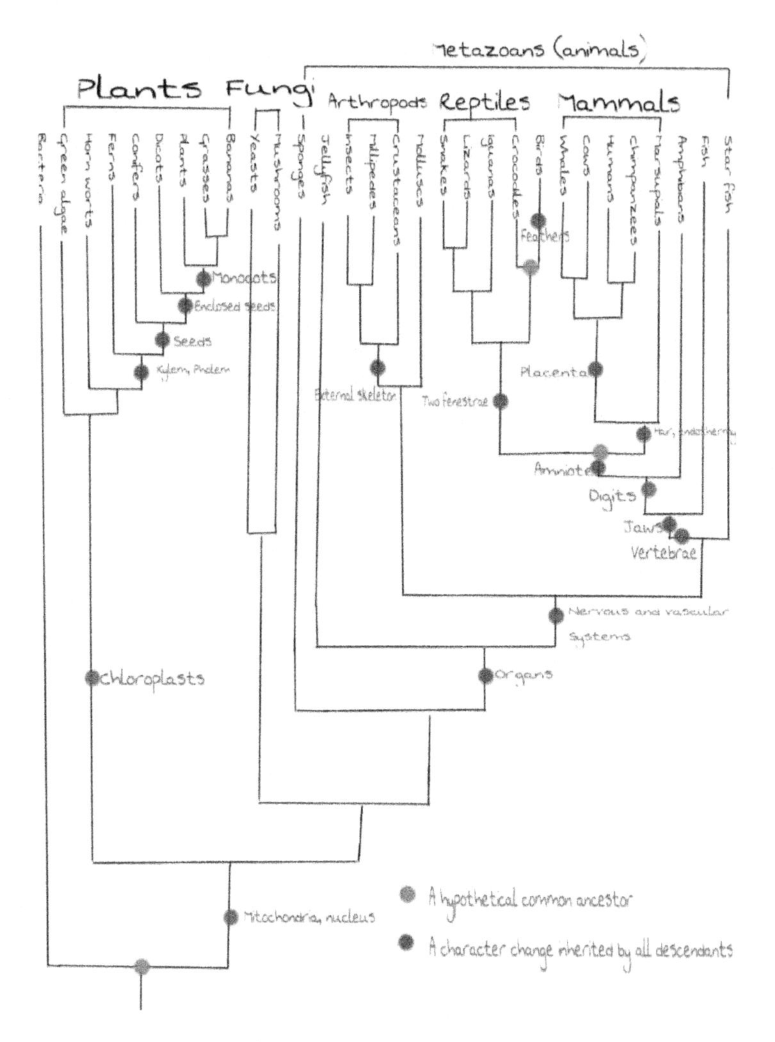

Fig. 10 Even more detailed evolutionary tree

Examination of fossil records in a timeline in Fig. 11 shows the changes that have occurred to the upper limbs of tetrapods. Tetrapods have four limbs and are a common ancestor of living amphibians, reptiles, birds and animals. From the diagram we can see how the 'fingers' changed with time. As the descendants of a common ancestor found new environments, a different structure of forelimb would be advantageous. This does not mean that all the ancestors would die out, only those species that found themselves in an environment they couldn't adapt to. As the environmental changes usually take many thousands of years, there is time to make the small changes needed to adapt to the new environments. For example, ice ages occur every 100,000 years. As the Earth cools, those creatures with fur or feathers that are able to keep themselves warm will be able to pass on their genes for fur or feathers better than those creatures that can't keep warm.

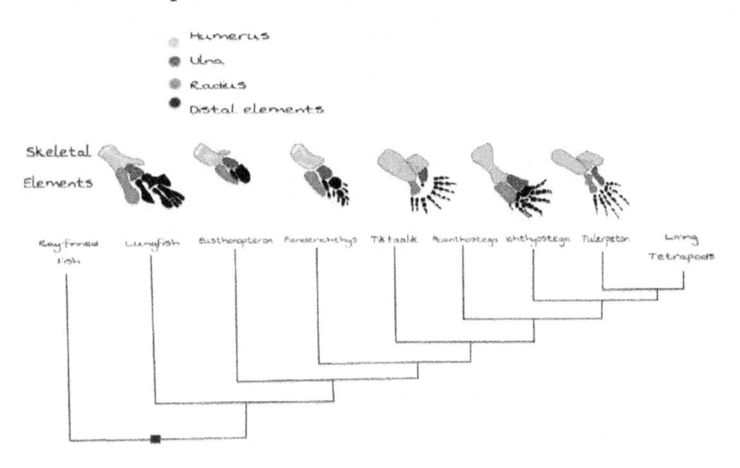

Fig. 11 Changes in tetrapod upper limbs over time

Analogous vs homologous

Structures that look similar but actually developed independently of each other are called analogous. This is true of bats' and birds' wings, as in Fig. 12. Although both are able to fly, the two wing types evolved independently of each other and at different times. Homologous characteristics are those that have a similar structure across different organisms and are evolved from a common ancestor.

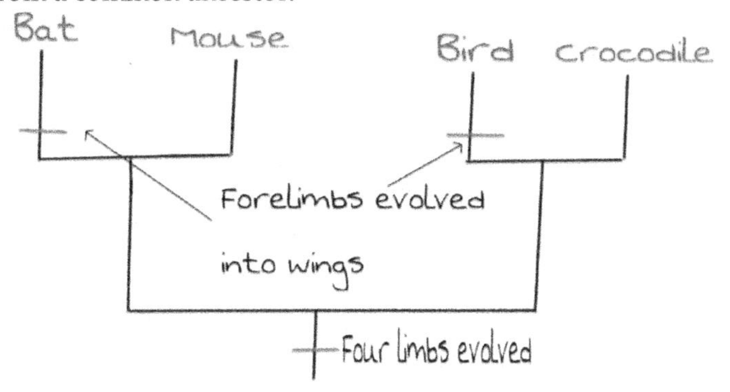

Fig. 12 Bats' and birds' wings evolved at different times and are analogous rather than homologous

Bats and birds both have four limbs and so are tetrapods. This is a homologous trait as it has only evolved once; therefore, all descendants with four limbs must be related.

One of the problems with the Linnaeus classification system is that organisms were grouped together by what they look like, or their 'phenotype'. This meant that originally bats were put in the same category as flying insects! All things with wings *could* be grouped together. Equally, all creatures living in water *could* be grouped together, so seaweed and whales would be in the same group. This obviously doesn't make sense: evolutionary relatedness cannot be determined by looks alone.

The eyes of humans, octopi and hummingbirds are analogous, as they evolved at different times. In fact, there are many structures that look and function in a similar manner to each other but are actually unrelated.

Darwin constructed his evolutionary tree based on what the creatures looked like - size and shape of finch beaks, for example. But there are limitations to grouping things together by phenotype, as we have seen. A new method of linking organisms by relatedness was needed. This couldn't occur until after DNA was discovered, along with the knowledge that genetic markers that were passed from generation to generation could be examined.

Who was LUCA?

If we return to Fig. 10 again and continue to trace the family tree back, we eventually reach point C. Found at the base of the evolutionary tree is LUCA. There may have been many cells before this one but, as they did not survive, they did not pass their genes on.

LUCA was a prokaryote and we know it would have had several key characteristics because all cells share them today:

- It had a lipid cell membrane.
- It used a nucleotide-based code - probably single-stranded RNA - to store information.
- It used a three base pair codon.
- It had ribosomes that produced proteins from the genetic information.
- It was able to divide itself to replicate.
- It probably maintained its internal concentration of sodium and potassium different to the outside concentrations by pumping the atoms in or out of the cell as needed.
- It used ATP for energy.
- Energy was generated by chemical synthesis.

As these traits are found in all cells today, by definition the ancestral cell must have had these traits too. From LUCA, and with several billions of years of time, all life around us has descended. This is known as descent with modification or a 'change in the characteristics of a population over time'.

Darwin is not the only way

There are other factors involved in evolution and determining which species survive. Darwin proposed natural selection - that is, descent with modification - but other aspects, such as migration and gene flow, genetic drift and mutation, are also involved.

Not all changes lead to reproductive advantages. The changes in genetic code between wet and dry ear waxes seen in Chapter 3 are considered to be neutral genetic changes.

Why can't an elephant breed with a mouse?

The answer is obviously size incompatibility, but there are other reasons why different species cannot breed with each other.

Genetic drift

Genetic drift refers to changes in the gene frequency due to random chance. This is not being pressured by environmental or adaptive changes. As it occurs by chance, it can lead to rapid changes over a short time span. An example is a meadow of grasses and flowers that were harvested or had pesticides applied. Rapid changes in the genes would occur as the plants and animals were lost. Equally, changes could happen due to weather and climactic events such as flooding or drought or warming of the seas causing coral bleaching, as we will see in Chapter 12. The pool of remaining genes has shrunk and future selection and evolution can only occur with the remaining genes. Once genes are lost, they are lost for good, as occurs with the poaching of wild animals to extinction.

Natural selection

Natural selection can be demonstrated by dark- and light-coloured moths. Prior to the Industrial Age, trees had lighter-coloured bark due to lichen. Light-coloured moths blended in with the lichen and thrived. The gene for a darker colouring was found in less than 0.001% of the moth population. However, during the Industrial Revolution, as pollution lay heavy and coloured the barks of trees, the dark-coloured moths had an advantage, as they were camouflaged against the dark, sooty trees. This resulted in the light-coloured moths being eaten by predators as they were more visible. The dark moths survived and were better able to pass on their genes, leading to them becoming the dominant colour in the population. By 1959, the dark genes were expressed in about 90% of the moth population. The Clean Air Act was enacted by the government of the day and started to change the pollution levels. The dark-coloured moths no longer had an advantage and so the frequency of genes in the two moth populations changed according to environmental pressures.

Mutation

This is the ultimate source of genetic variation. In plants and animals that reproduce sexually, there is a combining of the genes that also contributes to genetic variation. Some mutations involve the insertion, deletion or substitution of nucleotides, as discussed in Chapter 3. As previously seen, these changes can lead to no change, a small change, or a large change in overall phenotype.

Migration and gene flow

Geographic isolation has an effect on genetic variability. Imagine a population of lizards living on an island. Suddenly an earthquake tears the island in two. On each new island now live separate populations of the same species of lizards. Over time, the types of food available may vary as the climates are different - bananas may thrive on one island, while apples or mangoes thrive on the other. This means different genes need to be expressed to digest the different food sources. Changes to behaviour such as mating rituals may occur along with random genetic changes. Over many thousands of generations, the two populations will gradually change and diverge. If the two populations are now brought back together they may not be able to mate due to their physical and behavioural differences. This simple model results in two genetically different species that can no longer interbreed. Rather than islands and an earthquake, the same happens when hedgerows or woodland become isolated.

Gene flow reduces genetic variation. Imagine a garden 100m long populated with beetles. Beetles at one end may not be able to mate with the beetles at the far end. Although not strictly isolated, they never meet with each other due to distance and so never mate. However, if the beetles are able to mate with the beetles within 10m of them then, over many generations, there will be a graded change in gene frequency. This may or may not be enough to create a new species, but would alter gene frequencies between the beetles at either end of the garden.

Gene flow can also occur by dispersal of genes - for example, birds migrating. Birds migrating is gene flow, but if they had seeds or plant spores attached as they flew, the plants spreading to a new environment would be gene flow.

Speciation events

We can create new species ourselves! Dog breeding is a classic example. We see that by mixing different dog breeds together, new breeds or species can occur. Some would call these mongrel dogs. In 2016, the American Kennel Association recognised two new breeds: the American Hairless Terrier (in the terrier group) and the Sloughi (in the hound group). The same occurs with other animals - pigeons, horses or cattle, for example. Animals with characteristics or traits that are considered desirable are bred together with the aim of creating an even more desirable offspring. We can see the creation of a new species in a form of 'genetic modification'. Evolution can occur in a similar way except it takes many millions of years for each small trait to randomly occur and then accumulate to the point where there is a change in species.

Points A, B and C in Fig. 10 are speciation events. In terms of multicellular animals, the definition of a species is a group of individuals that can or do interbreed with each other. This definition cannot be applied to single-celled bacteria that divide by binary fission, nor can it apply if the organism breeds asexually. The lines are blurred as to whether the offspring should be called a hybrid or a mongrel or a new species. The boundaries can be difficult to define, but it is easier to see that a mouse and an elephant are different species both phenotypically (how they look) and genetically.

Summary

Evolution can be difficult to comprehend - especially where such huge timescales are concerned. Evolution doesn't just involve the theory of natural selection proposed by Darwin but includes responses to the changing environment. Throughout this book there are examples of modern-day evolution, ranging from HIV in Chapter 6 to antibiotic resistance and insecticide-resistant mosquitoes in Chapter 8 and ash dieback in Chapter 10.

Chapter 6 - HIV evolution and effects

HIV does not make people dangerous to know, so you can shake their hands and give them a hug:

Heaven knows they need it.

- Princess Diana

We have seen evolution over a long timescale in the last chapter, but organisms are still evolving today. An example of ongoing evolution is shown by a viral infection: HIV.

Human Immunodeficiency Virus (HIV) was first identified in 1981, since when it has killed more than 25 million people. This compares to about 25 million military deaths during the Second World War (from a total of 60 million deaths, including civilians and those dying of famine and disease). The majority of HIV deaths have occurred in sub-Saharan Africa, while about 450,000 UK citizens and 420,000 USA citizens have died. In 2014, there were 1.2 million deaths globally (613 in the UK) and 37 million people infected (103,700 in the UK). Nearly 16 million people are on treatment worldwide.

What is HIV?

HIV is a virus. Unlike the cells we have seen before, viruses cannot live on their own. They need to be housed inside a cell to be able to divide and replicate themselves. Viruses do not carry all the machinery they need to be able to produce their own proteins or copy their DNA. As they can only survive within a host cell, no host cell means no viral replication.

Superficially, HIV looks like a pretty simple particle, as shown in Fig. 1. The virus has three layers of protection:

an outer coat or lipid membrane; a protein outer capsid; and an internal protein capsid. Think of a box inside a box inside a box. The genetic code is locked inside the safety deposit box at the centre of the capsid, which prevents damage during its travels between cells.

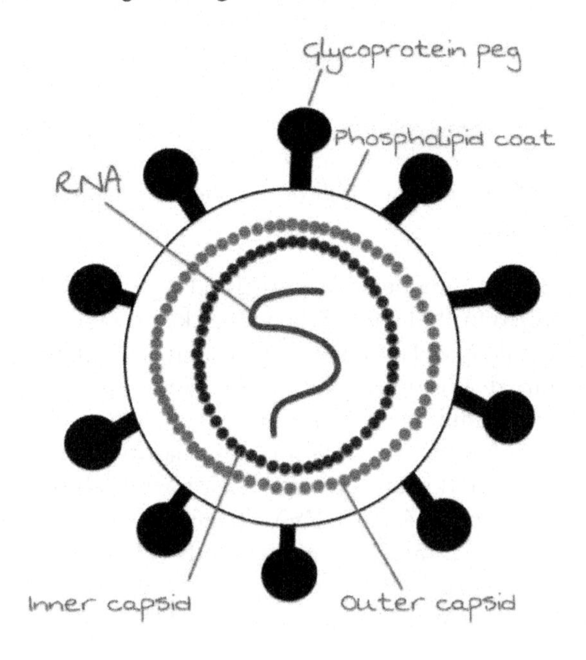

Fig. 1 Diagram of HIV virus

On the outer box there are two labels protruding from the surface that the virus particle uses to make sure it gets inside the right cells. The 'labels' are made from glycoproteins (gp), which are protein and sugar molecules joined together. These main two are named gp120 and gp41. We will see how these are used shortly.

The HIV genome carries only nine genes and consists of only 9,749 nucleotides. But instead of carrying the genes as DNA, HIV carries them as RNA in two single strands. Inside the central capsid are three proteins; these are enzymes ready to start reproducing the virus once it has gained access to a host cell. Once the virus reaches its destination, it wants to start replicating without waiting for new enzymes to be made.

HIV is a retrovirus; this means, when inside a cell, it is able to copy its own RNA into DNA and inserts this into the host's DNA. This is the reverse of the transcription we saw in Chapter 2. The machinery of the host cell then transcribes its DNA back into mRNA, but of course some of this copied DNA used to belong to HIV. The 'HIV mRNA' is then handed on to the ribosomes of the host cells, where it is converted or translated into proteins. These are, of course, HIV proteins. The enzymes from the HIV capsid now spring into action and cut the newly made proteins into the right shape. In effect, the virus hijacks the machinery of the host cell to make proteins and copies for itself. The HIV genome has been copied and assembled with their protein coats many thousands of times in each host cell. Each of these virus particles travels to infect other cells, replicates themselves again, and goes on to infect thousands more cells. And so the cycle of infection and replication continues.

More details

Of the nine genes in HIV, one codes for the proteins that are used in the internal capsule (known as 'Gag'); one codes for envelope proteins ('Env'); and a third codes for the trio of reverse transcriptase, integrase and HIV protease (named 'Pol', as in DNA polymerase). The other six genes are used in the replication and infection of the HIV and are named according to their function. Those regulating gene expression are Tat (trans-activator of transcription) and Rev, those trafficking proteins are

Viral Protein Unique (Vpu) and Negative Regulatory Factor (Nef) and finally Viral Protein R (Vpr) which helps the virus progress through the cell. The last gene is for Viral Infectivity Factor unsurprisingly named Vif.

The three enzymes inside the central capsid are reverse transcriptase, integrase and protease. As in Chapter 2, we found that DNA is transcribed into RNA, which is then used to create proteins by translation. HIV does something clever: it uses reverse transcriptase to turn the viral RNA into DNA. As its name suggests, integrase *integrates* the new DNA into the host cell. The protease from inside the viral capsid is used to trim the precursor proteins made by the host cell into the final mature HIV proteins.

How does HIV get into a host cell?

In a human body there is a complex network of cells involved in its defence. These are the white cells - so named to differentiate them from the red oxygen-carrying cells of the bloodstream. Different types of white cells are present: some kill on sight; others need pointing in the right direction; and others remember who the enemy is. These are shown in Fig. 2.

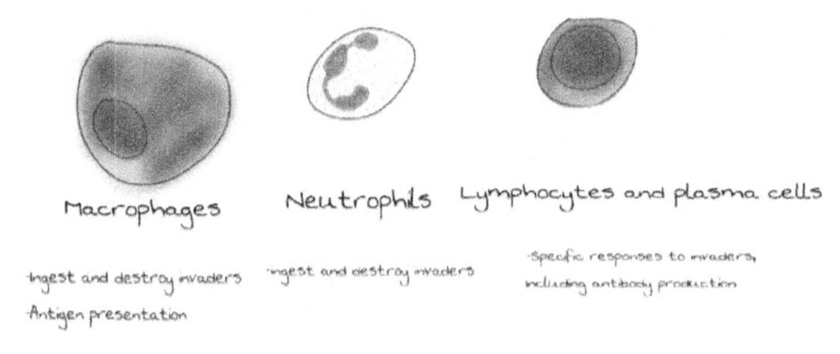

Fig. 2 Types of white cell

The cells prowling the blood looking for invaders are known as macrophages ('macro' = big; 'phage' = eater). Macrophages eat whatever they can find. To them, the body is like an all-you-can-eat buffet. They engulf and digest bits of cells, foreign substances, bacteria, cancer cells and anything they don't recognise. This process is called phagocytosis ('phago' = devour or eat; 'cyto' = cell; 'osis' = process). On all cells are proteins that stick out - akin to the labels on the surface of the outer box we met in Chapter 1. We met them earlier in this chapter on the surface of the HIV particle. These labels act as 'markers' to tell the macrophages which cells are friendly and which are not. If the macrophage can't find a recognisable friendly flag, it starts to eat.

If a bacterial cell gets into the body, the hunt is on. A game of cat and mouse ensues until the bacterium is caught by the macrophage. The bacterial cell is consumed and broken down into smaller constituent parts. Most of these parts are reused by the cell, but some protein fragments are used to help the immune system. These are little flags or tags called antigens; they are used to make bigger flags called antibodies. These bigger flags can be attached to other bacterial cells of the same type and are used to get the attention of other parts of the immune system. The production of antibodies is a second step of the response of the immune system and is the 'adaptive' or 'acquired' immune system.

The fragments of bacterial cell are transported through the macrophages to a special group of proteins on the cell surface known collectively as the major histocompatibility complex (MHC). The MHC is present on all 'antigen-presenting cells' such as macrophages. The process of bacteria breakdown and presentation on the cell wall is demonstrated in Fig. 3.

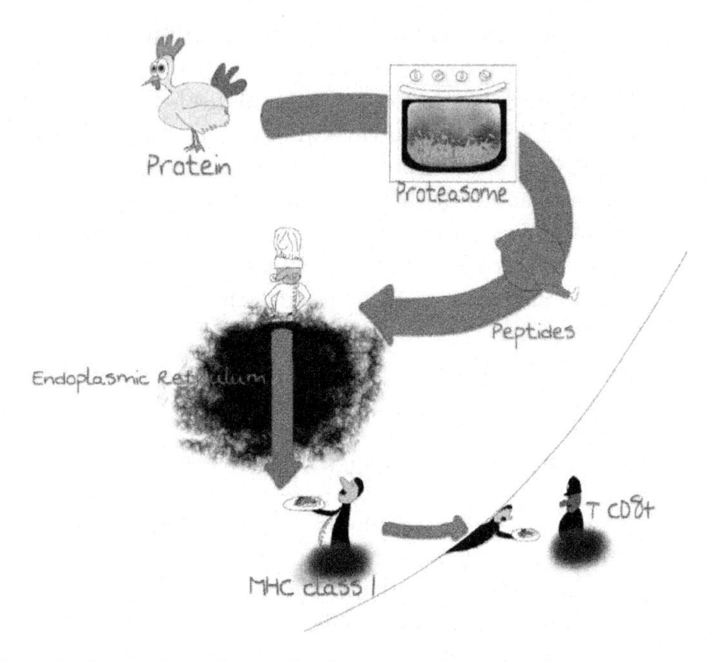

Fig. 3 Presentation of protein fragments by an antigen-presenting cell

The macrophage waves this unusual foreign protein fragment in the jaws of the MHC until a specialised cell comes along to help turn this antigen into an antibody. These helpful cells are known as T-helper cells. Macrophages use a T-cell receptor (TCR) to help dock with the T-helper cell, but need some help. The extra help comes in the form of a surface protein glycoprotein CD4, which 'pulls' the two cells close enough for the T-helper cell to get the information it needs. As a result, T-helper cells with these specialist glycoproteins are known as CD4+ T cells. There are many other receptors on T-cells to enable them to join to other cells and the CD4 receptor is on other cells too - including macrophages.

The antigen has been shown to the T-helper cells, but they can't produce antibodies without more help. Antibodies have been described as the 'memory molecule' of the immune system. T-helper cells in turn trigger another cell named a B cell to produce antibodies. As each fragment of protein being presented as an antigen is unique, each antibody produced is unique. Antibodies act like delinquent teenagers with cans of spray paint. They go around 'tagging' cells that they recognise as foreign, such as specific bacteria. This spray of paint alerts the authorities to investigate further. Rather than calling in the police, the body calls on cytotoxic ('cyto' = cell; 'toxic' = deadly) T-cells that recognise the sprayed 'tag' on the foreign cell and move in for the kill. This process is demonstrated in Fig. 4.

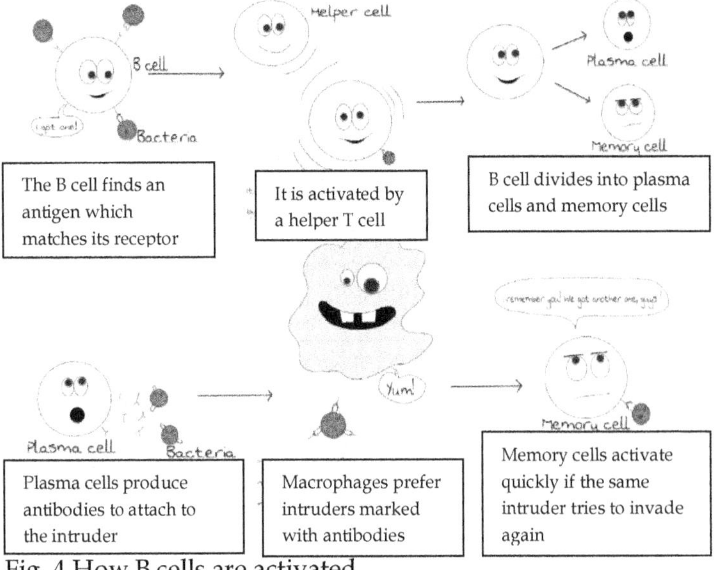

Fig. 4 How B cells are activated

Sometimes the body gets it wrong, however. If the antigens being presented are from the host's own cells, autoimmune diseases can occur, where the body attacks its own cells.

In a normally functioning immune system, bacteria are first digested by macrophages. The macrophages tell T-helper cells what to look out for. The T-helper cells tell a B-cell to make antibodies, so if the bacteria return there is a speedy mechanism for identifying the invaders and dealing with the infection more quickly the second time around.

How HIV fools the defence

We saw that HIV has an outer viral coat with surface markers on the lipid membrane; these markers are glycoproteins, labelled gp120 and gp41. The CD4 protein receptors on both the macrophages and T-helper cells recognise these as foreign. HIV uses this recognition to its advantage and becomes very adept at entering the white cells. The virus particle starts by fusing with the phospholipid bilayer. Fig. 5 shows the protein markers on the viral outer coat binding to the CD4 receptor. Once fused to the lipid bilayer of the host cell, the contents within the HIV protective coat are emptied into the host cell like a rotten egg breaking open and releasing its smelly contents into the surrounding kitchen.

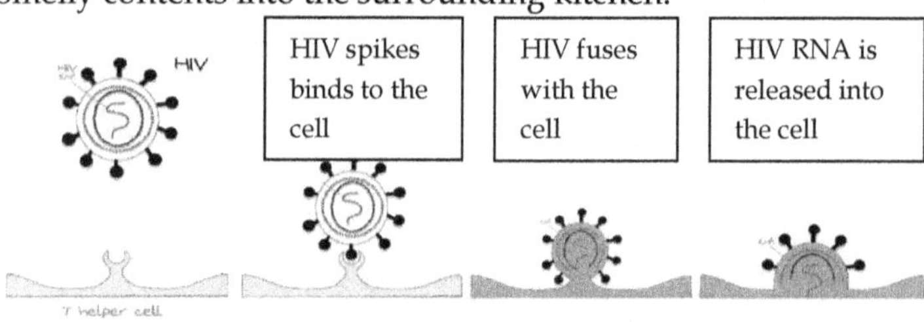

Fig. 5 Binding of HIV to a host cell

The three enzymes released from the capsid then get to work. First, an enzyme starts converting the viral RNA into double-stranded DNA. A second enzyme slices the host's DNA open and inserts this viral DNA. The host cell starts to produce many copies of viral mRNA. One piece of mRNA has the ribosomes making the proteins of the protective envelope, capsid and outer coat, and another arranges the proteins into the capsid itself. In effect, we have one piece of viral mRNA making Lego blocks and another building them together. Now the capsid proteins and copies of the viral RNA combine to form the internal capsid.

The viral surface proteins are sent to the surface of the host cell; this enables the HIV particle to leave the cell. As the capsid arrives at the surface and pushes through, it picks up a coating of the surface proteins. Budding of the viral particle occurs and it breaks free to go on to infect the next cell. Many thousand viral particles can be created from each host cell. The host cell dies in part because there are now foreign HIV proteins on the surface of the T-cell, and these are targets for the body's own cytotoxic cells. In effect, the immune system kills itself. The CD4 cells are forced to 'go rogue' as they are actively producing HIV virus and can also no longer divide, so become an ineffective member of the defensive immune system.

More details

First, the reverse transcriptase starts converting the viral RNA into double-stranded DNA. This creates a provirus which means it integrates into the host's DNA. By making use of the host's own RNA polymerase, two forms of viral RNA are created. One is called viral mRNA and recruits the host's ribosomes to start making proteins of the protective envelope, capsid and outer coat. The other is called viral progeny RNA and arranges the proteins into the capsid itself.

The surface glycoproteins gp41 and gp120 are made and sent to the surface of the macrophage. Budding occurs of the viral capsid through the lipid bilayer, creating a gp41 and gp120 studded protective coat.

Other mechanisms of cell death include the formation of proteins called death ligands causing cell suicide, and some of the HIV proteins being toxic to the host cells.

The phases of HIV infection

HIV infection develops in three phases: acute infection; chronic infection; and Acquired Immune Deficiency Syndrome (AIDS), as shown in Fig. 6. The initial acute infection can have very few symptoms. Some who are infected can feel as if they have a cold-like illness, but for many it is asymptomatic (symptomless). During this stage, the immune system is still functioning normally.

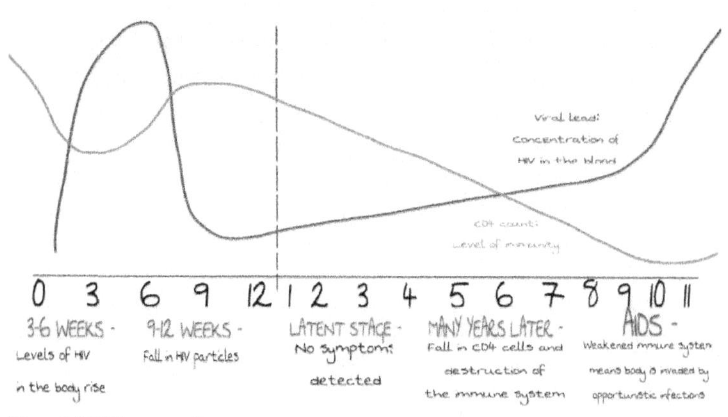

Fig. 6 HIV over time

As the number of viral particles increases in the host, the number of defensive CD4+ white cells gradually declines.

Blood vessels such as arteries, veins and capillaries are leaky. About 10% of the circulating volume of blood leaks out - but only the white cells and the liquid of blood known as lymph. These cells join the lymphatic system; this is a network similar to the circulatory system for blood but is used to specifically transport white cells and lymph back to the circulatory system. Along the way, lymph nodes are situated that act something like motorway service stations for white cells. Here, white cells can gather and wait until they are moved around to other parts of the body.

During the chronic infection stage of HIV infection, the lymph nodes are contaminated and seeded with viral particles. At this time, although the patient may not have any symptoms, they can still pass the virus on to other people. It can only be passed on through bodily fluids such as blood, semen, vaginal fluids or breast milk, but it can also be passed on during pregnancy and childbirth. Over many years of a chronic HIV infection, the immune system gradually weakens as the CD4 cells die off.

AIDS is said to have developed when symptoms of specific illnesses occur. These illnesses include opportunistic infections such as bacterial or fungal infections that would normally be managed by the immune system. More serious diseases such as Kaposi's sarcoma (a type of skin cancer) can develop. Towards the final stages of the illness, memory loss and coordination problems can occur. Altogether, without treatment, full-blown AIDS usually develops within eight years of infection followed by death within the following two years.

As HIV is a human condition, there are no other hosts that can be used to develop treatments. This makes finding cures and treatments potentially more difficult.

Mutation rates in HIV

As we saw in Chapter 2, DNA has a proofreading function, but RNA is more lackadaisical when it comes to checking itself. There is a much higher mutation rate when the reverse transcriptase forms DNA from RNA than when the DNA polymerase transcribes the DNA into (viral) mRNA. One research group suggests that reverse transcriptase makes a mistake once every 1,700 nucleotide bases. There are 9,749 bases in HIV. On average, every time the RNA is transcribed to DNA, five or six mistakes are made. As it is copied many thousands of times in any one cell, the chance of changes to the genetic code of HIV is very high indeed. Mistakes in the process lead to much genetic variability, and this variability leads to differing amounts of survival. If the changes are beneficial, the genes coding for the improvement will be passed on to their daughters and evolution will occur. Some of these changes will be harmful to the virus - for example, if the production of the capsid or the outer membrane becomes faulty. However, other changes will be beneficial - for example, if the glycoproteins are less visible to cytotoxic cells then the virus will survive for longer.

As HIV mutates rapidly it is hard to find a single viral target that would remain unchanged. As a result, there are many genetic variations between virus particles in every infected person.

Varieties of HIV

There are five varieties of HIV. The different strains vary in their infectiousness and lethality and can be split into two subgroups. HIV-1 is the largest group and affects 95% of those who have HIV. HIV-1 can be further subdivided into smaller groups. The most common strain of HIV is called HIV-1 Group M, which infects more than 33 million people. HIV-1 Group O numbers tens of thousands of victims; HIV-1 Group N has tens of people affected; and HIV-1 Group P has just two known infections. The smaller of the two main subgroups - HIV-2 - has one to two million infected people. It has similar symptoms of HIV-1 but is non-fatal.

Evolution of SIV and HIV

HIV is a human-only illness, but a similar virus called SIV (Simian Immunodeficiency Virus) exists in primates such as chimpanzees (*Pan troglodytes troglodytes*). There are marked similarities between HIV and SIV. Did five separate SIV infections cross to humans from great apes? If so, was this on five separate occasions or was it only once and HIV subsequently mutated within the human host into the different viral strains?

Within the great apes, SIV is found within at least 36 species. Each of these species has its own strain of SIV. For example, SIV in a chimpanzee is called HIVcpz, and in that of the Western gorilla is called SIVgor. SIV does not usually cause illness in its primate hosts. It appears that SIV has been present in the primate population for more than one million years.

Using genetic sequences from each of the different SIVs and HIVs, scientists can determine which SIV strains are more closely related to which HIV strains. Fig. 7 shows the analysis of the genes in a phylogenetic tree, revealing many similarities between HIV and different SIVs.

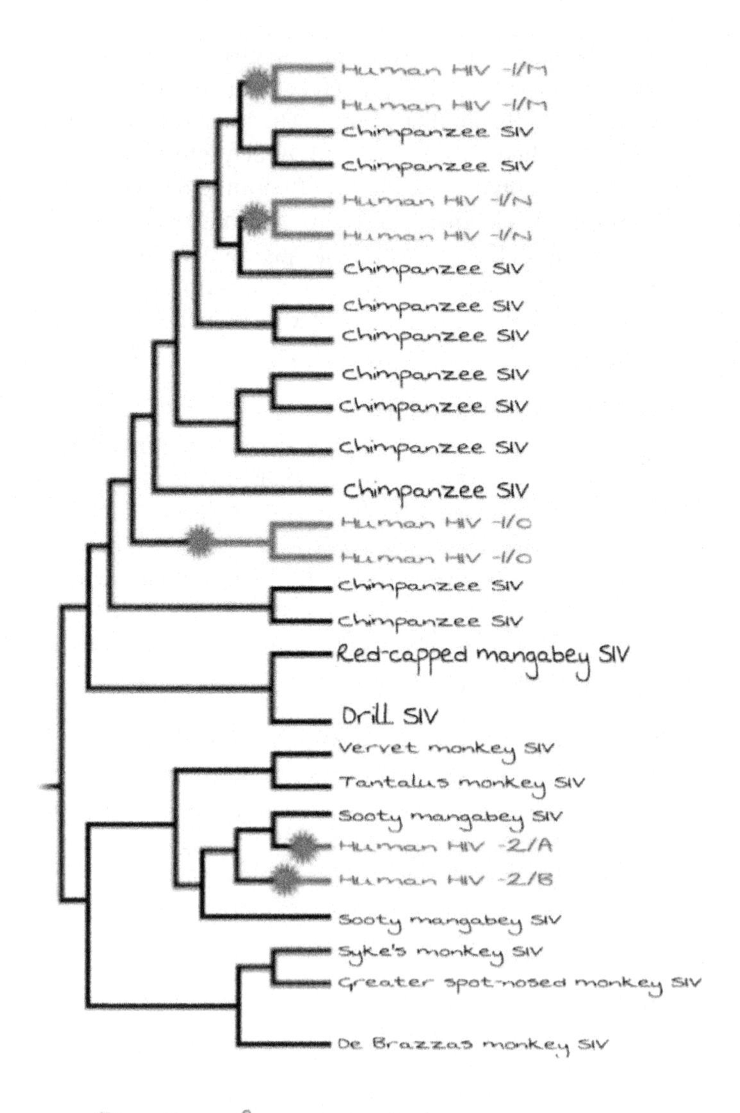

Fig. 7 Phylogenetic tree of HIV and SIV

The conclusion from the data is that on at least three (and probably four) occasions, HIV has been transmitted from primates to humans.

More details

The Pol gene is a core gene in HIV. It is approximately 3,011 nucleotide bases long and codes for reverse transcriptase, integrase and HIV protease. As these are coded on a single gene, all three proteins are made as one long protein, which is then cut by the viral protease into its individual enzymes.

The HIV could not function without these enzymes. By examining Pol genes within HIV and SIV, determinations about which SIV was related to which HIV were discovered.

Using the history of the Pol gene, it can be seen that there was great similarity between HIV-2 and SIVsmm from the sooty mangabey monkey. Strains HIV-1 M and HIV-1 N are similar to HIVcpz, and strain HIV-1 P is similar to SIVgor.

The information shown in Fig. 7 shows great overlap in the genetic sequences between the strains of HIV and SIV:

- SIVsmm (sooty mangabey monkeys) is very similar to HIV-2.
- SIVcpz (chimpanzee) is similar to strains HIV1 - M, N and O, but on analysis this cross between species has occurred several times.

How did SIV cross the species barrier?

Hunters in parts of West Africa can be exposed to bodily fluids during the capture, slaughter and cooking of bush meats. This may well have been when SIV crossed into humans as HIV. There is also evidence that monkeys kept as pets are another source of cross-species transfer.

Another virus has been discovered in primates that has crossed to humankind. The Simian Foamy Virus (SFV) was discovered in Cameroon in 1954. Like SIV, SFV is a retrovirus that also shows high levels of mutation in its genetic code. It has been isolated in many species including monkeys, cats and cows.

In 1971, a patient from Kenya was discovered to have Human Foamy Virus (HFV). Genetic analysis shows this to be similar to SFV isolated from chimpanzees, with up to 92% similarity in their base pairs. The local chimpanzee is *Pan troglodytes schweinfurthii* and lives in its natural habitat in Kenya. A study in 2007 showed that 24% of local people who had been bitten or scratched by primates had positive blood tests for SFV. The background population had a positive rate of 0.3%. This suggests that SFV is present in saliva and can pass to human contacts. SFV has also been found in workers in zoos and in primate centres.

Fortunately, unlike HIV, HFV and SFV are not dangerous or pathogenic. Human to human transmission has never been reported. Three of the genes (Gag, Pol and Env) are coded independently in SFV (unlike the single gene in SIV and HIV), which means they can be examined and synthesised separately. This allows HFV to be used for gene therapy, although there are some limitations.

Treatment of HIV

Increasingly, life expectancies are dramatically improving for people infected with HIV. The treatments are able to slow the progression of HIV to AIDS. Drugs have been developed that aim to interfere with different parts of the HIV life cycle.

When binding and fusion occurs, the HIV virus attaches itself to a T-helper cell.

- *The type of drugs that can stop this part of the process are called fusion or entry inhibitors.*

Once inside the cell, we have seen that HIV produces reverse transcriptase and integrase enzymes from the Pol gene. HIV can change its genetic material from HIV RNA into HIV DNA using these enzymes.

- *The types of drugs that can stop this part of the process are called nucleoside reverse transcriptase inhibitors (NRTIs), nucleotide reverse transcriptase inhibitors (NtRTIs) and integrase inhibitors.*

One drug that interferes with reverse transcriptase is azidothymidine (AZT), which was released in 1987. AZT produces the compound thymidine, which has a shape similar to thymine - the nucleotide used in the making of DNA. DNA needs an exposed hydroxyl group for the next base to bind to. Human cells don't use reverse transcriptase and AZT has a high affinity for the binding site on the HIV reverse transcriptase so is able to stop it working. AZT has a nitrogen group rather than a hydroxyl group, which means that no more bases can be added to the growing DNA polymer. AZT is like a pebble that gets into the cogs of a clock and gunges up the mechanisms, causing DNA prolongation to grind to a halt. Unfortunately, resistance to the drug can occur rapidly, as the gene for reverse transcriptase can mutate so that the shape of AZT no longer fits. The enzyme starts using thymine again. The development of AZT resistance is shown in Fig. 8.

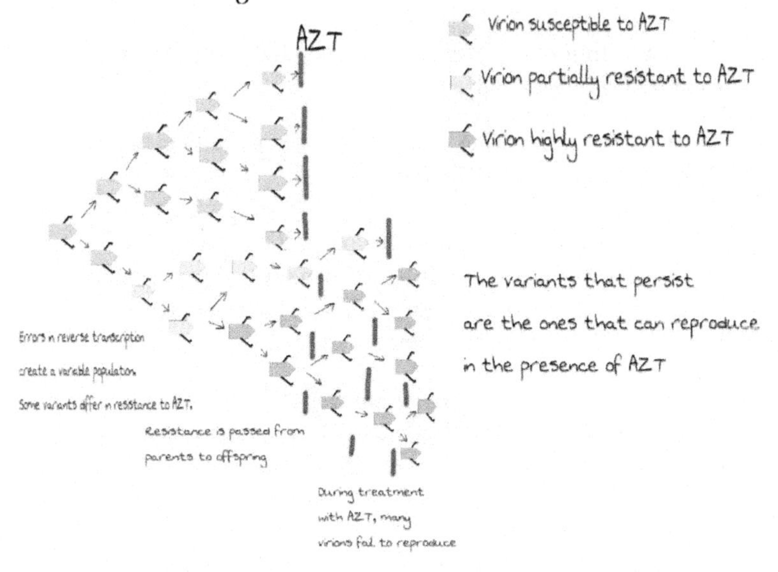

Fig. 8 Evolution of AZT resistance

Further enzymes are used during the translation of mRNA. The proteins produced are transported towards the edge of the cell, and are then used for producing more HIV. Copies of HIV are assembled and packaged into their capsids and outer lipid coats prior to budding and maturation. These combine to form new HIV particles, which are then released from the T-helper cell and are ready to infect other cells and begin the process all over again.

- *The types of drugs that can stop this part of the process are called protease inhibitors.*

Due to the high likelihood of developing resistance, these drugs are often used in combination. This strategy is called Highly Active Antiretroviral Therapy (HAART).

Summary

As humankind continues to encroach on parts of the world that have previously been unexplored, it is likely that we will discover new illnesses that we have never been in contact with before. Great apes had been living far away from humans until the last few decades, but destruction of their habitats has brought them into closer vicinity. This means the opportunity for passing a virus from one species to another will only just have been realised.

Chapter 7 - Bacteria and virulence

Don't blow it - good planets are hard to find.
– Time *magazine*

Bacteria are everywhere. Every part of the Earth has bacteria - from the frozen poles to the equator, from the top of Mount Everest to the depths of the oceans. Bacteria have found a way to survive and thrive in many different environments.

Bacteria are essential to life on Earth. American microbiologist Carl Woese, whose work we look at in Chapter 9, remarked: '*It's clear to me that if you wiped all multicellular life-forms off the face of the earth, microbial life might shift a tiny bit, but if microbial life were to disappear, that would be it – instant death for the planet.*'

Bacteria carry out many functions that are essential to the continued existence of multicellular life on the planet. Bacteria themselves have only one aim: to make more bacteria. For the Earth, they provide nutrients in the form of carbon, nitrogen, phosphorus and others that other organisms can then make use of. Some decompose the dead or break down food into smaller molecules.

What makes bacteria?

Bacteria are single-celled prokaryotes. They all have a phospholipid bilayer, sometimes called the plasma membrane. Some have one further protective layer made of peptidoglycan, while others have two protective layers: a peptidoglycan layer and an outer membrane made of fats or lipids. The peptidoglycan layer is a mesh made from long polymers of alternating amino acids (peptide-) and sugars (-glycan) molecules. These polymers are cross-linked by covalent bonds (a strong bond where atoms share electrons) to form a net-like structure. There can be many layers of peptidoglycan laying one on top of another to form the cell wall.

One way to classify bacteria is by physical characteristics such as whether or not they have the outer lipid membrane. In 1884, Danish bacteriologist Hans Christian Gram invented the Gram staining technique, where the peptidoglycan layer is stained. He used two dyes, crystal violet and safranin, which showed the different bacteria as purple or pink under the microscope. If the bacteria show as purple, they are said to be Gram positive, as they only have a peptidoglycan layer. If the bacteria are pink, they are said to be Gram negative and a second protective layer has a phospholipid layer; i.e. the dye cannot reach the peptidoglycan layer. This is shown in Fig. 1.

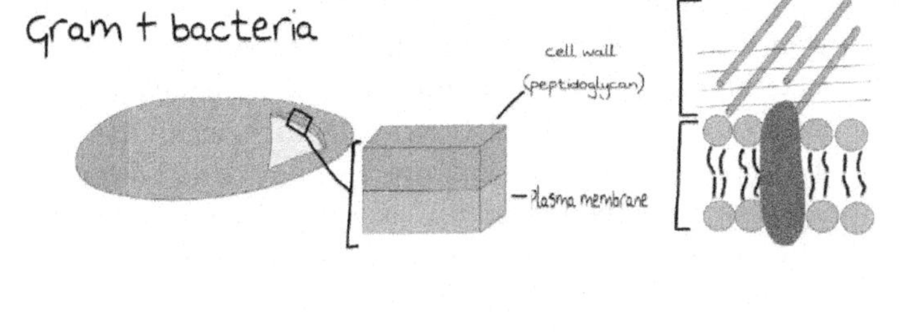

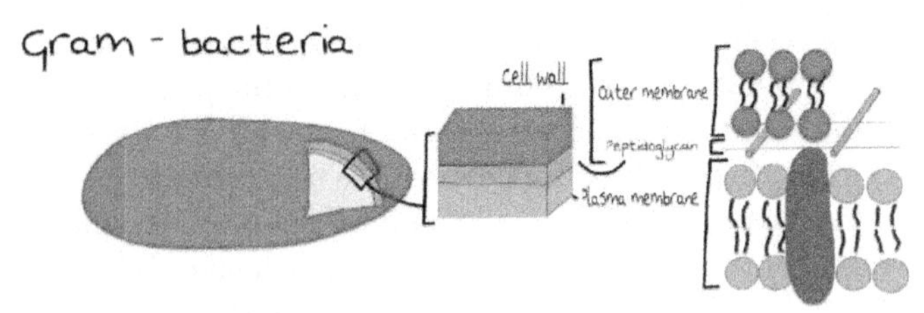

Fig. 1 Differences in cell walls

Knowing whether or not bacteria have a peptidoglycan layer is vitally important, especially if the bacteria are pathogenic (disease-causing). For example, as we will see in Chapter 8, antibiotics aim to disrupt key parts of the bacterium or its life cycle while leaving the host unharmed. In the next chapter we will see how penicillin works by preventing the peptidoglycan layer from forming and subsequently how resistance develops.

Another physical characteristic to sort or classify bacteria is by shape. Bacilli are rod-shaped; cocci are round or spherical. Some are spiral-shaped and others are irregular in shape.

Size matters

Bacteria are small in comparison with the larger eukaryotic cells. While eukaryotes can measure 10-100μm, bacteria are in the order of 2μm. 1μm is the equivalent of 0.001mm. As mentioned, there are no specific structures or internal organelles inside a bacterium like there are in eukaryotes, but there is some organisation within the cell. In comparison with eukaryotes, bacterial DNA is smaller and is usually circular and not wrapped around spools or histones. Bacteria possess ribosomes for making proteins, but these too are smaller. However, cell replication is far faster - in part because there is less DNA to copy. As discussed in Chapter 4, bacteria replicate themselves by dividing and splitting into two by a process called binary fission.

The genome is the entire DNA of a cell, but in a bacterium not all the DNA is stored in one place. There is an extra store of information in a separate circle of DNA called a plasmid. The plasmid is able to exchange genetic material with other bacteria or the surrounding environment, and genes can be gained or lost in this way. This is known as horizontal gene transfer (HGT) and can be important in passing on antibiotic resistance between bacteria. HGT refers to passing genes from one mature cell to another rather than vertical gene transfer (VGT), which occurs during replication and cell division. HGT can also be used for removing genes from the cell if mistakes have been made during replication of the DNA.

More details

The number of nucleotide base pairs ranges from about 1×10^5 (or 1 followed by five 0s; i.e. 100,000) to 1×10^6 in prokaryotes and from 1×10^7 to 1×10^{11} in eukaryotes. The number of base pairs reflects the number of genes; e.g. ranging from 552 genes in the smallest known genome of *Nanoarchaeum equitans* to 4,485 genes in *E. coli*. Comparatively, eukaryotes can have more than 30,000 genes in wheat or 23,000 genes in humans.

There are very few non-coding regions in the DNA of bacteria, meaning a far higher coding density. This implies a more compact and efficient storage system with less retention of superfluous DNA. For example, in *E. coli*, over 90% of the genome codes for proteins, which means an average gene density of one gene per kilobase (kb). One kb is 1,000 bases. In a bacterium, 1,000 bases can make you a protein. In eukaryotes the additional introns mean the effective coding density is far lower than in prokaryotes. For example, in a human the average coding percentage is less than 5%, meaning a gene density of close to one gene per 30kb.

How do we know bacteria are there?

Most bacteria have never been seen and more than 95% have never been grown in the laboratory, so how do scientists even know they exist?

Analysis of a special gene within each and every bacterium provides proof of the variety of species around us. We know that some genes are essential for life - whether as bacteria or as complex organisms. We saw in Chapter 3 that there are three essential molecules needed for a cell: DNA, RNA and proteins. Ribosomes make the proteins that each cell needs, so the DNA coding for the ribosomes itself is considered essential to life. Without it, a cell couldn't produce ribosomes, couldn't produce proteins and so couldn't survive. For bacteria to grow, a working copy of the gene is needed; this is the 16S rRNA gene. We saw that ribosomes are made of large and small subunits and the 16S rRNA codes for the smaller bacterial subunit. The 16S rRNA gene is said to be 'conserved' - that is, that a working copy must exist in every cell. It is only 1,542 bases long and the very small size means it is cheap and easy to copy in the laboratory. There are small differences between the 16S rRNA gene between organisms that, when put together on an evolutionary tree, create the diagram we saw in Chapter 5. It is the small differences in the sequence of the nucleotide bases that determine which bacteria are being examined. Chapter 14 discusses the background behind the techniques used, such as PCR.

It is by examining the 16S rRNA in, for example, a soil sample, then comparing the genes discovered, that allows scientists to determine which bacteria are present. This information forms the basis of Fig. 2.

Bacteria: the planet's most prolific species

In Chapter 5 we saw an evolutionary tree. Fig 2 shows an even more complex tree, from Nature in 2016 and is used to demonstrate the vast variety within bacterial species. The top half of the figure shows bacteria, whereas eukaryotes - all the multicellular plants and animals we see around us - are relegated to a thin sliver of green in the bottom right corner.

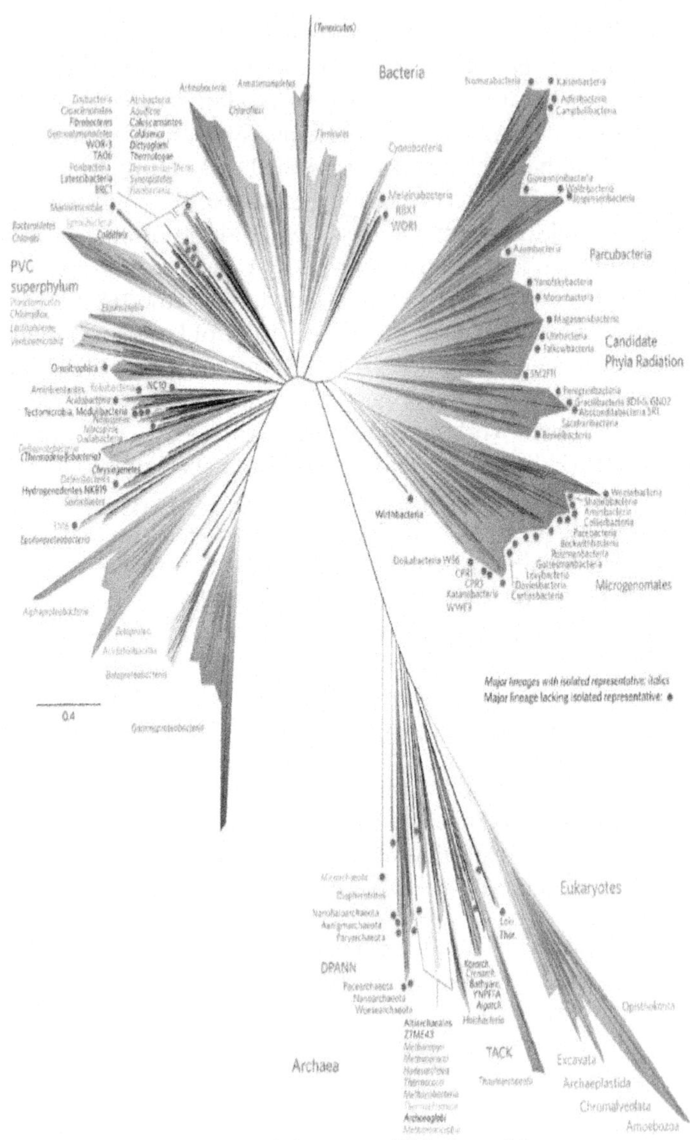

Fig. 2 Demonstration of the tree of life from Nature.

Extremophile bacteria

More organisms live within a relatively narrow temperature range. Water freezes at 0°C – if a cucumber is put into the freezer and then defrosted, it turns into a mushy mix of cells. The cells of the cucumber burst open as the ice crystals form, which destroys the cells' structure. When thawed, and with no cell wall to hold the contents together, it turns into an unstructured mess. However, some plants and animals, such as penguins, have found ways to survive in the frozen wastes of Antarctica (average temperature -49°C). Other animals have found ways to thrive in hot, dry places such as deserts. The hottest recorded temperature on land was 56.7°C in 2013 in Death Valley in America. More than 50 types of mammal, 300 species of birds and reptiles and even fish have been found to live there. All these animals and plants have remarkable adaptations to the cold or heat. Some bacteria, though, have an even greater temperature range that they can survive, from -80°C to more than 100°C. Such bacteria are known as extremophiles.

Bacteria live in diverse environments in which conditions and resources are continually fluctuating. To thrive in such a challenging situation, bacteria must be able to respond quickly to environmental stimuli. They must be able to use a variety of resources and quickly switch between alternative metabolic pathways by activating the required set of genes and repressing unwanted ones, as we saw in Chapter 3.

Bacteria have been found in cold dark lakes buried half a mile down under Antarctica or at the bottom of the deepest part of the Ocean in the Mariana Trench. Some examples of marine extremophiles are looked at in Chapter 12.

Different extremophiles are tolerant of different conditions. For example, acidophiles and alkaliphiles are organisms that grow when the pH is less than three or greater than nine respectively. An aerobe needs oxygen to live, whereas an anaerobe doesn't. A thermophile is comfortable between 45°C and 122°C; a hyperthermophile lives optimally at temperatures over 80°C; and a cryophile reproduces at temperatures lower than -15°C. An osmophile is capable of growth in a high sugar environment, whereas a halophile likes salty conditions. A barophile or piezophile enjoys being under pressure, such as in the ocean depths.

Polyextremophiles have more than one extremophile characteristic. For example, a thermoacidophile manages best if the temperature is high and the pH low.

The field of astrobiology is interested in extremophiles. NASA has shown that some bacteria can thrive in very high gravity environments or in low gravity environments such as when taken on the space shuttle. *Salmonella* was grown on board and showed more virulence than Earth-grown *Salmonella*. *Pseudomonas* created different biofilms when on the space station that could have implications for antibiotic resistance.

Living in acid: Helicobacter pylori

It was believed that nothing could live, survive or thrive in highly acidic environments until the extremophiles were discovered. One such highly acidic place is the human stomach. The stomach produces hydrochloric acid and the pH is maintained between 1.5 and 3.5 depending on whether or not we have eaten. The acid acts as a defence mechanism to prevent harmful bacteria entering the intestines. It also activates enzymes such as pepsin, which break down the proteins in our food.

For many years, it was thought that stomach ulcers were caused by a combination of rich food, too much alcohol and stress. One amazing discovery turned that thinking upside down. In 1982, two Australian doctors and scientists not only discovered bacteria living in the stomach, but also linked it to the cause of stomach ulcers. This bacterium was *Helicobacter pylori*. It is Gram negative and a curved rod shape under the microscope.

Dr Marshall and Dr Warren investigated the causes of stomach problems. After struggling to infect piglets with *H. pylori*, Dr Marshall had an endoscopy (where the inside of a stomach is examined with a camera on a flexible tube) and deliberately infected himself with the bacteria. Within only three days he had developed stomach problems. After eight days he had another endoscopy and saw his stomach was very inflamed – a condition called acute gastritis. Samples were taken that showed *H. pylori* growing. He took a course of antibiotics after 14 days and eliminated the bacteria, allowing the ulcer to heal. In 2005, the doctors won the Nobel Prize for Physiology or Medicine for their discovery.

H. pylori infects about 50% of the world's population. Most people do not suffer from symptoms, but there is a lifetime risk of 10–20% developing stomach ulcers and 1–2% developing stomach cancer.

H. pylori shows an affinity for acidic conditions. Could it have evolved to be a useful part of the human bowel flora? Could removal of the bacteria cause problems? Some vitamins and minerals – for example, iron - are better absorbed in acidic conditions. Iron can exist in different states - as Fe^{2+} or Fe^{3+}. In acidic conditions, it reverts to Fe^{3+} and is better absorbed. Could a lower stomach pH reduce the iron absorbed and if so, what effect might this have? The body needs iron in the production of red blood cells, so lower iron uptake would mean more likelihood of iron-deficient anaemia. However, some bacteria also require iron and if deprived will reproduce more slowly. This adds another layer of complexity to the balance between whether a higher or lower pH of the stomach is more important.

What are bacteria good for?

Throughout this book we can see examples of what bacteria are able to do. In Chapter 8 we delve into the relationship between humans and bacteria. Chapter 10 discusses the importance of bacteria in the nitrogen cycle. Chapter 11 examines the bacteria living in the soil, and Chapter 12 shows bacteria in the marine environment.

Waterborne bacteria

The first ever cells on the planet lived in water, so it is unsurprising that many bacteria live and thrive in water today. A huge variety of microorganisms live in water, including viruses, bacteria and protozoa, with bacteria recognised as the largest group.

A balance between different species is achieved through competition for resources and space. Fortunately, most pathogenic bacteria are poor competitors and are usually wiped out by less dangerous bacteria.

Some of the most common bacteria causing disease are *Shigella, Salmonella* and *E. coli*. *Shigella* causes dysentery (a type of severe diarrhoea often with blood present). It is most commonly found in food but is also found in water in developing countries. *Salmonella* causes gastrointestinal disease with diarrhoea. It is most prevalent during the summer months and is carried by about 15% of farm animals, which is a major reservoir for infection. Foods such as fruit and vegetables can become contaminated if they have been in contact with livestock, manure or untreated water. Infection can occur from contact with individuals with diarrhoea or from unwell animals. *E. coli* causes gastrointestinal disease and urinary tract infections. Cholera also causes diarrhoea and is discussed in detail in the next chapter.

The variety of species of bacteria in water is high but fortunately the incidence of disease is low. There is no single method available to identify all species and testing for each individually is expensive. However, there is a way to work out which water might be contaminated and that is to use *E. coli*.

Escherichia coli

E. coli is a type of bacteria known as a coliform. These are Gram negative and rod-shaped, and are one of several different coliform species that are present in faeces. The relationship is shown in Fig 3. Most coliforms do not usually cause serious illness.

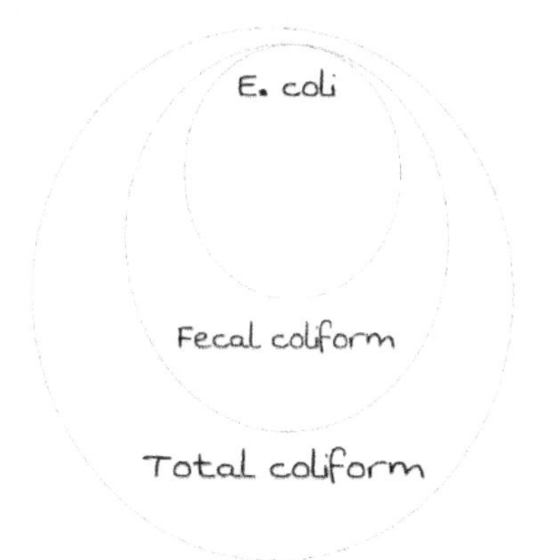

Fig. 3 Relationship between *E. coli* and coliforms

However, some strains of *E. coli* cause travellers' diarrhoea and are a major cause of diarrhoeal disease in the developing world, especially among children. People usually contract *E. coli* through ingestion of water contaminated with human or animal faeces. Symptoms of *E. coli* gastroenteritis include watery or bloody diarrhoea with associated cramps and low-grade or no fever.

Of the many different types of *E. coli* only a few are harmful to human health. In testing water quality, *E. coli* is used as 'indicator' bacteria. This means if they can't grow in the water it is unlikely that other pathogenic bacteria could grow there. Therefore, by using a harmless strain of *E. coli* it is easy to identify clean water.

Faeces from animals can be problematic if it contaminates the water supply. In New York City alone, the 500,000 owned dogs contribute 68,000 kilograms of faeces and 340,000 litres of urine *every day*. In rural areas, farm animals and wildlife contribute large quantities of waste material to potential water sources. The average number of bacteria *per gram* of faeces for a horse is 12,600, in a cow is 230,000, and in a chicken is an amazing 1.3 million. Sheep have a staggering 16 million bacteria per gram, but this is small compared to the 23 million from a dog. For comparison, humans have 'only' 13 million per gram.

Virulence factors

Bacteria have many different ways of getting inside a host to cause a damaging infection. These are known as the virulence factors that allow bacteria to gain a foothold. These can include substances that allow the bacteria to stick to the host tissue called adhesins. Invasins are molecules that allow the bacteria to invade the host cells. Molecules that impede the host's defence mechanisms are called impedins. Aggressins cause direct damage to the host such as by toxin release, while modulins are able to modify the behaviour of the host cell.

Broadly, bacteria use a variety of virulence factors in different ways; they can be characterised as follows:

Acquire. This refers to the acquisition of genes coding for more virulence factors. These genes can be gained by HGT- the transfer of DNA from a plasmid to another bacterial cell, as seen earlier. Alternatively, bacteria can pick up genes from the surrounding environment in a process called transformation. In transduction, a virus helps to transfer DNA from one bacterium to another, and in transconjugation one cell donates DNA to a recipient cell via a hair-like projection called a pilus (plural = pili). A pilus is used to help one bacterium 'stick' to another to pass on information; it looks a bit like a jet plane being refuelled mid-flight.

Activate. The activation and regulation of genes is important to coordinate an attack on a host. The control of gene expression was discussed in Chapter 3. Some bacteria only release toxins when there is a large enough number or a high enough concentration of them. If a bacterium makes itself known to the immune system too soon, it can be wiped out before an attack even begins.

Adapt. We have seen bacteria surviving in many different environments; they have also become adept at managing many 'stresses'. Just as Chapter 3 showed *E. coli* using different food types, other bacteria have genes to control protein production when it is too acidic. *H. pylori*, mentioned above, produces an enzyme that generates ammonia from urea to allow it to adapt to low pH conditions. This gene is upregulated by being in the stressful situation of bathing in a pool of acid.

Advance. *Listeria monocytogens* causes listeria. It is able to live and reproduce within a host's cell. It makes a 'tail' (sometimes called a comet tail or actin rocket) from a protein called actin, which enables it to move and advance through a host cell and out the other side to infect the next cell.

Alight. To alight or exit the host is important so the species of bacteria can find a new host to infect. Some bacteria are transmitted by the faecal-oral route. The bacteria come out in faeces and, if hands aren't washed, the bacteria will end up going back into the mouth at some point, for example, when eating. Good hand-washing hygiene is very important to reduce the risk of bacterial spread, especially before mealtimes or during a time of diarrhoeal illness.

Alter. Some bacteria alter their own cells to make entry more effective. One type of *E. coli* rearranges part of its cell wall to form pedestals that help it gain access.

Appropriate. Iron is an important element for bacteria to grow. Some bacteria switch on specific genes to appropriate or seize iron supplies from its host. These genes make proteins that bind to iron and are controlled in a similar way to the gene expression we saw in Chapter 3. In *E. coli*, they are regulated by the Fur gene.

Attach. To reduce the risk of the bacteria being removed, it needs to 'stick' or attach itself to the host cell. It can do this by creating a biofilm or 'adhesins', as we will see in Chapter 8. Small pili-like projections called fimbriae (singular = fimbria) allow bacteria to stick to surfaces.

However, if the genes coding for fimbriae are lost then the mutant bacteria can no longer stick and will no longer hang around long enough to cause a disease.

Attack. Many bacteria release toxins to cause damage to the host, such as cholera, as explored in the next chapter.

Avoid. *Streptococcus pneumoniae* has evolved a capsule that allows it to avoid the host's immune system and avoid being recognised. *Legionella pneumophila* manages to avoid the immune system by taking up residence inside a white cell called a phagocyte. It can survive without being digested.

Many bacteria use several virulence factors simultaneously. Cholera is a good example, as we will see in the next chapter. Salmonella is another bacterium that uses multiple factors. These Gram-negative bacteria is often found in warm-blooded farm animals. Different species of salmonella cause food poisoning, typhoid fever or paratyphoid fever. Salmonella is killed by heat, so cooking chicken or eggs thoroughly is a sensible idea. What virulence factors do they have to allow them to thrive in a hostile host? To start with, they can lose the genes for their 'tail' (this type of tail is called a flagellum), as the immune system targets this tail as a foreign body. Losing its tail means the salmonella can avoid the defensive white cells. They attach to host cells by producing adhesin molecules and have fimbriae. They attack by producing cytotoxins that damage the host cells and cause diarrhoea. Another toxin is produced from the surface of the bacteria called an endotoxin and causes the host to suffer a fever. The bacteria avoid the immune system by having a capsule to avoid recognition and produce a chemical that means they are less likely to be devoured.

Overall, bacterial virulence is complex and multifaceted.

Bacteria causing disease in plants

Some species of bacteria are pathogenic towards plants. In 1878, fireblight was discovered affecting apple and pear trees in America. This was caused by *Erwinia amylovora*, which limits the trees' growth. Corn can wilt when infected by strains of *Erwinia stewartii* and potato ring rot is caused by *Clavibacter michiganensis*. The latter can survive on machinery and packaging material to reinfect more potatoes at a later date. These bacteria can easily be spread by wind or rain, by insects or animals or by infected farm equipment. As we saw earlier, some microorganisms prefer particular conditions to thrive in. Some prefer temperatures under 22°C, such as *Pseudomonas syringae*, and others over 22°C, such as *Xanthomonas campestris*. Both of these affect the growth of bean plants; if a bean plant is infected by both simultaneously, you can imagine how it would struggle to reach its potential.

Crop rotation plays a role in protecting against disease. If a field of potatoes has been affected by potato ring rot it would be foolish to plant potatoes in the same field the following year. Rotation to, say, beans or carrots means the *Clavibacter* bacteria can no longer thrive. Some farming techniques reduce biodiversity such as using a single monocrop year after year; this increases the likelihood of bacterial infections affecting the harvest in subsequent years. Some bacteria are beneficial to farmers, as we will see in Chapter 11, and there is an important balance between maintaining biodiversity, conservation and farming.

Summary

The importance of bacteria to the planet cannot be overemphasised. Without bacteria, many processes would cease to function. The variety of types is overwhelming and the scale can be hard to imagine. Although bacteria are only there to pass their genes on to the next generation, they have evolved different ways to affect plant and animal hosts - and cause a lot of damage to some along the way.

Chapter 8 - Bacteria and humans - friends or foes?

In nature, there are neither rewards nor punishments; there are consequences.

- R. Ingersoll

As we have discovered, life and bacteria are everywhere. This is true for all the parts of the human body that bacteria can reach. Numerically, the body contains about the same number of human cells as bacterial cells - somewhere between the staggering 10 million and enormously staggering 100 trillion. Because bacterial cells are far smaller than human cells, they take up far less space and we don't see them individually. Collections of bacteria - for example, when an infection takes hold - can be seen as pus or discharge. Collectively, the bacterial species inhabiting our bodies are known as the human microbiome. We will peer a little closer at a few aspects.

Bacteria on the skin

Our hands come into contact with many surfaces through the day and as a result are in contact with, and carry, a lot of bacteria. The main bacteria found on our hands are *Streptococcus, Staphylococcus* and *Lactobacillus*. One study showed that one in ten bank cards and one in seven banknotes carried faecal bacteria. In 2009, another study showed that over one-quarter of bus and train commuters had faecal bacteria on their hands. This adds to the million or so bacteria *per square centimetre* on each hand, representing about 150 different species. Washing one's hands after going to the toilet doesn't seem to happen as often as it should!

The palm of the hand has more bacteria than the fingertips. The scalp can also have a million bacteria per square centimetre. The armpit may only have 10,000, while the number found on the abdomen and forearms is in the thousands. Overall, more than 4,700 different species of bacteria have been found to live happily on the skin. Generally, women have more bacteria than men. Surprisingly, in only 17% of people is there the same bacterial species on the left hand as on the right hand. Hand washing does remove some species of bacteria, but other species actually *increase* in number and are more abundant after washing. It is still not clear which bacteria offer a protective effect and which may contribute to skin problems such as eczema, acne or psoriasis, although it is likely that the vast majority are neutral. A study showed that skin bacteria do not change much over our lifespan.

Will 'bacterial fingerprinting' ever be used by crime fighters? It is possible, although there are some difficulties. About 45% of those hands examined had rare bacteria present and these could be used to identify who had been at a crime scene. Scientists examined office workers' hands and could identify who worked at which computer by using the patterns of bacteria. Bacteria can hang around for up to two weeks and don't rely upon an unsmudged fingerprint for identification. At the moment, this technique is only 70–90% accurate, so not yet good enough for a courtroom.

Bacteria in the mouth

Although there are more than 700 strains of bacteria in the mouth, two are particularly important for tooth and gum disease. *Streptococcus mutans* enjoys the same sugary food we do. It produces a waste product that is acidic, and it is these acids that cause tooth decay in humans. The bacteria produce a biofilm that allows them to stick to the surfaces in the mouth. A biofilm is a slimy glue-like substance and in the mouth gives a yellowish colour to the teeth. Brushing helps to break up the biofilm. Run your tongue across your teeth and see if you feel a biofilm - if you do, brush your teeth! There is more on biofilms later in this chapter. Brushing or chewing sugar-free gum also reduces acidity and protects the teeth. *Porphyromonas gingivalis* does not usually live in a healthy mouth, but when it is present it causes periodontitis - damage to the tissue and bones around the teeth, causing them to fall out.

The rest of the bacteria are spread across six phyla and include *Firmicutes, Actinobacteria, Proteobacteria, Bacteroidetes, Fusobacteria* and the TM7 phylum. Most of these can't be grown or cultured in the laboratory but have been identified using the 16S rRNA gene we saw in Chapter 7. It is not clear which bacteria may be protective, but more pathogenic bacteria are being identified. One day, it may be possible to swill your mouth out with protective bacteria that control the harmful ones!

The links between poor dentition and other health problems have long been recognised. For example, it is thought that chronic infection by bacteria such as *Porphyromonas gingivalis, Actinobacillus actinomycetemcomitans, Treponema denticola* and *Tannerella forsythia* in the mouth leads to some types of heart disease. Patients with periodontitis have a higher risk of heart attacks and strokes. Rheumatoid arthritis and periodontal disease have a similar mechanism of chronic inflammation, while the DNA from the bacteria causing periodontal disease has been found in the joints of some patients with rheumatoid arthritis.

A study in 2016 showed the link between oral bacteria and pancreatic cancer. Once again, *Porphyromonas gingivalis* is implicated, as is a species that can also cause periodontal disease - *Aggregatibacter actinomycetemcomitans*. This mix of these two showed the risk of pancreatic cancer increased by 50%.

Rather than the presence of bacteria just affecting humans, human activity also affects the bacteria in the mouth. We have seen that brushing the biofilm away disrupts the ability for bacteria to stick to teeth. There are dramatic changes in the levels of different species of bacteria in a smoker's mouth. Smoking also increases the acidity of the saliva in the mouth. It was shown that smoking increased 150 bacterial species but decreased another 70 species. Proteobacteria make up about 12% of the bacteria in a non-smoker's mouth, but this fell to 5% in a smoker's and these bacteria help the breakdown of some toxins and chemicals. The amount of *S. mutants* in a smoker rose and we have seen this contributing to more tooth decay. Fortunately, the bacteria in the mouth return to the levels of that of a non-smoker if the smoker quits. It may be in part due to changes in oral flora of bacteria that smokers develop more oral cancers.

Bacteria and mood

The interplay between bacteria and mood is becoming recognised. In an experiment, scientists took bacteria from the bowels of stressed and anxious mice and exposed non-stressed mice to the same bacteria. After a few weeks, these mice too were showing signs of stress and depression. In another experiment, rats exposed to the bacteria *Toxoplasma gondii* became fearless around cats. A study in 2012 showed that cat owners exposed to the same bacteria had a higher risk of suicide.

Bacteria and obesity

Our gut bacteria are able to break down some chemicals that we cannot. We have seen that bacteria can release compounds such as alcohol from yeast or acid from mouth bacteria. Some gut bacteria such as *Firmicutes* can produce waste that we can use as energy, to increase our absorption of fat and break down fibre. Changing our diet can have profound effects on our microflora. Even a ten-day diet of junk food is enough to wipe out 40% of the species in our bowels. The high sugar and refined grains decay the health of our bacterial biome. A course of antibiotics can have the same effect. One reason diets may fail is because of the loss of microflora diversity. The overdominance of *Firmicutes* leads to a less diverse bowel - specifically with a lower number of *Bacteroidetes* species. If we no longer have the balance of bacteria, we may get more calories out of our food than the next person with a more favourable balance. Therefore, bacterial imbalance can drive obesity, but obesity can also drive bacterial imbalance.

How do harmful bacteria affect our bodies?

Over the last century, death rates in the Western world due to infectious disease have fallen. Whereas tuberculosis, pneumonia and diarrhoea caused over half of all deaths in 1900, this proportion fell to about 7% in 2000. In developing countries, the death rate from infections remains similar to that of Europe and America 100 years ago.

Bacteria found on and in our bodies naturally can end up causing disease through different means. Some only cause disease when the host is weakened; others only when they can gain entry, e.g. through an injury or surgery. Some bacteria cause no problems until they obtain extra DNA that causes them to develop 'super powers' – or at least gain virulence factors. Other bacteria, such as tuberculosis, are never found on our bodies unless they are causing an infection.

Several things need to occur for bacteria to enter the body. First, the host needs exposure. This could be when they consume contaminated food or water or breathe in bacteria from the air. Using a tissue or handkerchief when sneezing or coughing can reduce the spread of infectious droplets. The bacteria need to stick or adhere to the skin or the cells lining the mouth, airways, stomach or bowels. They can then invade either between the host cells or into the cells, which causes colonisation and growth. As they do this, further local toxic effects can occur or spread to other parts of the body. These combine to cause cell damage and disease.

Some bacteria are happy to live outside the cell, others inside. They can form pathogenic (one species benefits, one suffers); commensal (one benefits, the other neither benefits or suffers); or mutualistic (both species benefit) relationships.

Cholera

Cholera is caused by *Vibrio cholera*. It was found in London in 1832 at a time when the disease was thought to be spread by a bad smell or 'miasma'. Dr John Snow was a local doctor who carefully plotted the cases of a cholera outbreak on a map of the city. He thought it was being spread in the water supply and managed to trace the source to a water pump handle in Broad Street. He removed the handle of the pump and the cases of cholera fell immediately. Unfortunately, he died in 1858 prior to the acceptance of his ideas in the 1860s. His findings helped inspire Joseph Bazalgette to provide the sewerage system used in London today.

Cholera was originally found in the River Ganges in India. It was transmitted around the world by travel and exploration during the 17th century. Cholera remains a problem in countries with inadequate hygiene, poor water and sewage systems and those water supplies contaminated with *V. cholera*. Water-treatment plants can be damaged after a natural disaster such as flooding or tornadoes. Annually, cholera causes more than 100,000 deaths worldwide and can kill within hours of infecting a host.

More details

The cholera bacterium uses an adhesin molecule to adhere to the outside of the intestinal cells of the host. It causes damage by producing an exotoxin - the name for a toxin excreted by a bacterium. The toxin has two subparts: A and B. As the B subunit sticks to receptors on the intestinal cell, part A is activated and is able to enter the cell. This causes several things to happen. First, sodium ions, which are normally reabsorbed from the bowel into the cells, are blocked. Second, to balance the ions in the body, chloride ions are lost to the bowel, followed by bicarbonate and water. This combination leads to a lot of salt and water in the bowels; as it can't be reabsorbed, it is lost from the body as profuse diarrhoea. Cholera uses several virulence factors that we met in the last chapter, including attaching, attacking (toxin production) and alighting. When the number of cholera bacteria has increased and reached a certain density, a gene is switched on that codes for a protein that promotes their release. Cholera is jettisoned from the bowel wall and leaves the host in the diarrhoea. By escaping the dying host, it returns to the water supply to live in a biofilm until it can enter another victim.

Prevention is through clean water supplies - especially with chlorination - and good hygiene, although there are vaccinations available. Treatment is with oral rehydration and salt replacement - that is, sugars and salts mixed in water that is drunk. The importance of rehydration treatment is shown in the reduction in death rates from about 40% to about 1%.

Tuberculosis

Tuberculosis (TB) is caused by *Mycoplasma tuberculosis.* In 460BC, during the Greek empire, Hippocrates labelled this disease as 'consumption'. Egyptian mummies from as long ago as 2400BC have evidence of TB in their spines. A similar bacterium, *Mycoplasma bovis,* affects cattle and may have been transmitted to humans via infected milk perhaps around 5000BC. In 1839, the term 'tuberculosis' was coined, derived from the Latin *tuberculum,* 'small swelling, bump or pimple'. Nobel Prize Winner Dr Koch found the causative bacteria in 1882.

TB is a serious condition that can affect many parts of the body but commonly infects the lungs. It is estimated that globally 1.5 million people die each year from TB; over the whole of human history this number is as high as one billion.

The symptoms include a persistent cough - especially with blood in the phlegm - weight loss and sweating at night. It can be transmitted to others through droplets in the air during coughing. Most healthy people have immune systems that can remove it from the body. However, in some people the bacteria avoids the immune system and remains in the body without causing symptoms. This is known as 'latent TB' and is almost like Sleeping Beauty in a deep sleep or dormant phase. About one-third of the human population has latent TB; in around 10% of these the TB will become active at some point in their lifetime. If the immune system weakens and loses control, the bacteria start dividing and spreading, becoming 'active TB' and causing symptoms.

Diagnosis is made by growing the bacteria from the phlegm if the lungs are affected, or by X-rays of the chest. If the TB is elsewhere in the body, small samples called biopsies are taken for analysis. A skin test called the Mantoux test involves injecting a small amount of protein from dead TB under the skin. If the body has a reaction and the area becomes red and swollen it means the body has previously made antibodies to TB during a previous contact. This could happen in someone who had latent TB, previously had active TB, or a small reaction can occur if they have been vaccinated.

Prevention is by Bacille Calmette-Guérin (BCG) vaccination. This is a weakened strain of TB bacteria; antibodies are formed that build up immunity and enable the body to fight the 'real' *M. tuberculosis* bacteria if the person comes into contact with it. The vaccination gives up to 80% of people protection against TB, but cannot prevent everyone who comes into contact with the bacteria from becoming infected.

Treatment is with antibiotics. Problems with antibiotic-resistant TB mean that at least two antibiotics are given together. Antibiotic resistance is discussed shortly. The course of treatment runs for many months. One side effect of the antibiotic rifampicin is that urine can turn red and look like blood. This looks scary but is harmless. Problems are occurring in part due to the emergence of drug-resistant *M. tuberculosis*; in part due to lack of compliance in completing the courses of antibiotics; and also the return of *M. bovis* to infect humans.

Malaria

Diseases can be caused by microorganisms other than bacteria. One example is malaria, which is caused by the parasite *Plasmodium* - a single-celled organism transmitted by the female mosquito to the human host. Malaria uses an 'insect vector'; i.e. the causative organism is transmitted by an insect. The mosquito is from the genus *Anopheles*, of which there are more than 400 species. There are many types of *Plasmodium* parasites, but only four cause malaria in humans: *P. falciparum* (the most dangerous), *P. vivax*, *P. ovale* and *P. malariae*. Malaria is commonly found in tropical countries, especially near stagnant water – the breeding grounds for mosquitoes. The estimated number of cases varies from 200 million to 365 million; between half a million and two million deaths occur annually. 90% of deaths occur in sub-Saharan Africa, for which *P. falciparum* is responsible. *P. vivax* is the dominant malaria parasite in most countries outside of sub-Saharan Africa.

Transmission is complicated due to the life cycle of the parasite, which differs in the mosquito and the host. Mosquitoes lay their eggs in water, where their larvae hatch and emerge as mosquitoes. The females need to nurture their eggs and do so by having a meal of blood. The mosquito bites and *Plasmodium* parasites enter the host, where they travel to the liver and start to multiply. When in sufficient number, the parasite returns to the bloodstream to invade red blood cells. Here they continue to multiply, eventually causing the cells to burst open and spill their contents back into the bloodstream. This takes about 48 hours and can lead to a big drop in the number of whole red blood cells. Anaemia is the loss of red blood cells; specifically, this is haemolytic ('haem' = blood; 'lytic' = bursting open or rupture) anaemia. The parasite roams free in the bloodstream and infects the next mosquito to bite the diseased person. When in the mosquito, the parasite continues to replicate itself; it migrates to the salivary glands to be transmitted in the next bite to the next host, and so the cycle continues.

Diagnosis is based on examining a sample of blood under the microscope. If red cells that have burst open and the parasite itself are seen, then treatment begins in earnest. Newer, more accurate, techniques such as using PCR to multiply and amplify specific segments of the genetic code are used. There is more on PCR in Chapter 14.

Prevention is aimed at preventing mosquito bites, using insect repellents and insecticides, mosquito nets and antimalarial drugs. Different combinations of antimalarials are needed in different countries depending on the main parasite found there. There can be side effects from the different treatments, but none are as bad as the death caused by malaria itself. A vaccine against *P. falciparum* is in development and was undergoing trials in 2016. The results should be known by 2020 or 2021.

'Vector control' means removing the mosquitoes. The inside of a house is sprayed and this is effective at keeping mosquitoes away for 3–6 months in 80% of the houses sprayed.

Dichloro-diphenyl-trichloroethane (DDT) was first made in 1874 and was developed into an insecticide in the 1940s. Initially, it was enthusiastically welcomed for crops, homes, gardens, and in the fight against malaria. By 1970, though, it was banned in America due to concerns about environmental and toxic effects to humans and animals. It is currently listed as a 'probable cancer-causing agent' or carcinogen. DDT hangs around in the environment for a long time and accumulates in the fatty tissues of the body. However, the World Health Organization (WHO) supported its use for indoor spraying in 2006 as they felt the risks of malaria were worse than the health risks of DDT. In some parts of the world DDT is still used for crop spraying.

DDT-resistant mosquitoes were first recorded in 1959 in India. The resistance developed in just a few months rather than years and demonstrates rapid evolution. Even if 99% of the mosquito population are killed, the resistant 1% that remain are able to breed and repopulate the area quickly. The resistance is due to the production of a specific detoxifying protein called CYP6Z1. This is one in a class of enzymes known as cytochrome P450 monooxygenases. Humans have cytochrome P450 proteins produced by our livers to break down toxins. For mosquitoes to produce this protein they need the knockdown resistance gene (KDR). A mosquito without the gene is said to be 'KDR minus'. The difference between resistance and susceptibility to DDT is due to a substitution or switching of just one single amino acid for another. This substitution has huge implications for the structure of the final protein produced and, vitally, the function of the enzyme.

Treatment uses a drug called quinine from the bark of the cinchona tree. It tends to be given in tablet form, but in severe cases of malaria it needs to be given directly into the bloodstream. Transfusions of blood are sometimes needed if the anaemia is severe.

Just as there has been a resistance developed to DDT, some *Plasmodium* is becoming resistant to antimalarial regimes. The first drug resistance was to chloroquine in the 1960s, which had been the best and most widely used drug for preventing malaria.

Changing climate is likely to see the spread of mosquitoes to countries not previously affected. If Northern Europe continues to see rising temperatures, mosquitoes will find a home and malaria may become a threat in previously uninfected countries.

Smallpox

Although we have not yet eliminated malaria, the eradication of smallpox has been achieved.

Diseases can be caused by viruses, as we saw in Chapter 6 with HIV. Smallpox is caused by one of two viruses in the *poxviridae* family: *Variola major* or *Variola minor. Variola* means 'spotted' or 'pimple'. Originally, this disease had been known as the 'pox' or the 'red plague', but eventually became known as smallpox to distinguish it from the 'great pox' of syphilis. Smallpox had been around for the last 10,000 years until the last known case in 1977. *V. major* was the more severe infection and caused a spectacular rash and death in 35% of cases, while those who survived were scarred, deformed or blinded. By the 18th century, smallpox was killing 400,000 Europeans annually and caused one-third of all incidents of blindness. Even in the 20th century, it killed up to 500 million people worldwide.

The highly contagious virus responsible was passed through droplets in the air from the coughing or sneezing from an infected person. It was not transmitted by insects or animals.

Prevention of smallpox came about by a curious route. Prior to Dr Jenner in 1796, variolation was used to try to prevent the risk of death from smallpox. Variolation involved scratching the skin of a healthy person with a skin scab from someone with a milder form. This was unpleasant and often caused serious side effects, including death.

In 1796, Dr Edward Jenner (who had suffered from his own variolation) noted that milkmaids who had had a milder disease called cowpox did not contract smallpox. Cowpox is caused by a different pox virus and is an untroubling illness. Jenner hypothesised that a cowpox infection conferred some protection or immunity from smallpox. He took samples from Sarah, a dairymaid who had the infected sores of cowpox, and deliberately spread them to the son of a gardener - an eight-year-old called James Phipps. As expected, James had a mild illness but recovered within the week. Dr Jenner recognised that cowpox could be transmitted from person to person, but would cowpox really protect against smallpox? Bravely, he infected James with smallpox. If he was wrong about the protection afforded by cowpox, James could have died. To everyone's great relief (especially James'), James did not develop smallpox. Dr Jenner published his findings in 1798 and subsequently performed more experiments into inoculating more people and confirming his original findings. This was the first vaccine; the word root *vacca* comes from the Latin for cow.

By 1803, export of the smallpox vaccine for mass vaccination had begun. In 1813, the US Congress passed the Vaccine Act to ensure access to all.

A worldwide mass immunisation program against smallpox led to the last natural case being in 1977 and a laboratory accident causing a case in 1978. Now the virus only exists in two laboratories in the world - one in America, the other in Russia - solely for research purposes.

Antibiotics

Antibiotics have been important in reducing illnesses and deaths caused by bacteria. If asked which group prescribes the most antibiotics in the world, it would be reasonable to guess hospital doctors or GPs. Worldwide though, most antibiotics are used in the farming and agriculture sector, as they boost the weight of cattle and poultry prior to slaughter.

Antibiotics can be classified in different ways depending on how they do their job. If they can kill bacteria, they are known as bactericidal ('cidal' = to kill). If they stop them from replicating and multiplying, they are bacteriostatic ('static' = to stop). Some also affect the translation or transcription of bacterial DNA into proteins.

It is not only antibiotics that are bactericidal. Bleach and disinfectants kill bacteria, but cause so much damage to all cells that the hosts would be severely affected too. Antibiotics labelled as bactericidal do not kill 100% of the bacteria, but are closer to 90–99% effective; they let the immune system do the rest. Some antibiotics have a bactericidal effect on some bacteria but a bacteriostatic effect on others. Antibiotic chloramphenicol is bactericidal against *Streptococcus pneumoniae* but bacteriostatic against *Staphylococcus aureus*.

Penicillin is a bactericidal antibiotic and is discussed below. A disadvantage of some bactericidal antibiotics is the effect that killing large number of bacteria quickly has on the host. Bacterial death causes their cell contents to be released as they lyse or burst open. These cell fragments are spread around the body and can cause further damage. When meningitis is caused by *Streptococcus pneumoniae*, the deaths of the bacterial cells cause the release of chemicals that contribute to even more inflammation of the brain and an increased death rate.

Bacteriostatic antibiotics stop the rapid replication of bacteria, thus frustrating their ability to reach high numbers. Antibiotic tetracycline blocks the bacterial ribosomes and stops new proteins from being made. They do not interfere with protein synthesis in the cytoplasm of the eukaryotes due to differences in the translation enzymes we met in Chapter 3. Some bacteria use the same trick, but against the host. Cycloheximide is made by the bacteria *Streptomyces griseus*; it only inhibits protein synthesis in eukaryotes and does not affect protein synthesis in bacteria. Another antibiotic, rifampicin, inhibits the RNA polymerase of bacteria but not in the eukaryotic nucleus and is used in the treatment of TB.

The first antibiotic and penicillin

The first antibiotic purified and synthesised was not penicillin but arsphenamine. It was synthesised in 1911 and was effective against only one disease: syphilis. This very limited range of action meant the hunt was on to find an antibiotic that worked against a much larger range of bacteria. These would be the 'broad-spectrum' antibiotics we have today. In the 1930s, more compounds were discovered, the first of which was a sulphonamide (brand name Prontosil) created at Bayer laboratories by Gerhard Domagk in 1932. He successfully treated his own daughter and saved her from having her arm amputated when she developed a skin infection called cellulitis from an embroidery needle. This discovery earned Domagk a Nobel Prize in 1939.

Unfortunately, the Nazis made him refuse the prize and he was arrested by the Gestapo and detained for a week.

Alexander Fleming discovered penicillin in 1928 and started treating infections in 1942. Penicillin comes from a mould called *Penicillium glaucum*. The same mould is used in blue cheeses such as roquefort or stilton. It was known in the 1800s that bacteria would not grow in the presence of this mould, but the species could not be identified. There are tales that in the Middle Ages mouldy bread was used to prevent infections on wounds that soldiers had received in battle.

The *Penicillium* fungus has more than 300 species. It is usually found in the soil and prefers cooler climates. It is the cause of some of the moulds on fruits, bulbs and even garlic. Some species are harmful to animals such as mosquitoes and Vietnamese rats, while other species cause damage to lubricants and machinery such as fuels, oils and lubricants.

Penicillin is actually a group of antibiotics that includes penicillin G for intravenous use and penicillin V for oral use. It works by preventing the peptidoglycan cell wall that we saw in Chapter 7 from forming cross-links. The lack of a cell wall means these bacteria would show purple when Gram stained. Gram-positive bacteria include *Staphylococci*, *Streptococci*, *Clostridium* and *Listeria.* With the cell wall destroyed, salts and water enter the bacteria cell, causing it to swell up and burst. Molecules such as DNA and RNA and other valuable proteins leak away into the surrounding environment. Penicillin is mostly effective against bacteria that don't have an outer membrane protecting their peptidoglycan layer. As viruses are structurally different from bacteria and do not have the peptidoglycan layer or the same chemical pathways, antibiotics do not affect viruses.

Where do antibiotics come from?

The major groups of antibiotics are macrolides, cephalosporins, fluoroquinolones, sulphonamides, tetracyclines and aminoglycosides. Each group has several antibiotics within it.

We have seen that penicillin, the first broad-spectrum antibiotic, was discovered from a mould. Where have the other major groups been discovered?

Macrolides are broad-spectrum, bacteriostatic antibiotics that were discovered in the soil bacteria *Streptomyces erythraeus* in the 1950s.

Cephalosporins are broad-spectrum antibiotics derived from the fungus *Cephalosporium acremonium* found in 1945 in a Sardinian sewage outfall pipe.

Fluoroquinolones are broad-spectrum antibiotics found accidentally in 1962. Nalidixic acid was found as a by-product during the process of making the antimalaria drug chloroquine. This led to the discovery of a group of antibiotics called fluoroquinolones.

Sulphonamides such as Prontosil mentioned above started out as industrial dyes which were found to stain cell walls.

Tetracyclines were first produced by the bacteria *Streptomyces,* found mainly in the soil and decaying material. It works by blocking bacterial protein synthesis.

Aminoglycosides are derived from another soil-dwelling bacterium, *Streptomyces griseus* (which also makes cycloheximide, as seen above).

As we can see, all antibiotics bar the sulphonamides and fluoroquinolones were discovered from other organisms - mainly soil bacteria and fungi. If biodiversity is lost, the developments that have taken place in bacteria and fungi through evolution will also be lost.

What is antibiotic resistance?

Imagine you take some drug to kill a particular species of bacteria that is causing you harm; for example, as a skin, urinary or chest infection. The drug may kill most of the bacteria, but due to genetic variation a small proportion of these bacteria may differ enough from the rest. This is represented in Fig. 4. Any slight difference may make them fit enough to survive the drug; for example, a slightly thicker cell membrane. The vacancy left by the killed bacteria allows the surviving bacteria to make use of the food and space left behind. They can develop and reproduce more rapidly without the competition of their neighbours. At this point, the surviving bacteria in your body will pass on the genes for resistance to their daughters and almost all will be fit enough to survive the drug. As a result, the drug becomes ineffective: this is called drug resistance. If antibiotics are used, the weakest bacteria are wiped out, leaving the strongest to survive. The strongest are those with the most resistant genes to the antibiotics. If a different antibiotic is used next time, the weakest remaining bacteria are again removed from the population, leaving the even stronger/most resistant ones to reproduce and divide.

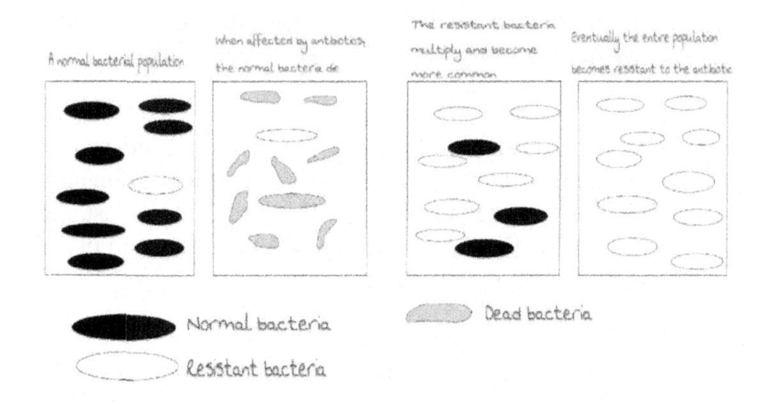

Fig. 4 Antibiotics killing bacteria and creating a resistance strain

Mutations in the bacteria may result in resistance to antibiotics, and this resistance is passed to subsequent generations by vertical gene transmission. However, resistance may also be passed from one species to another by horizontal gene transfer, which involves the passing of plasmids of DNA in bacteria, as we saw in Chapter 2.

More details

Let's consider the development of penicillin-resistant bacteria. We saw that Gram-positive bacteria form a cell wall from peptidoglycan sheets, and penicillin stops these forming during reproduction. The penicillin molecule contains a B-lactam ring (Fig. 5), which blocks the two walls making polymers binding together.

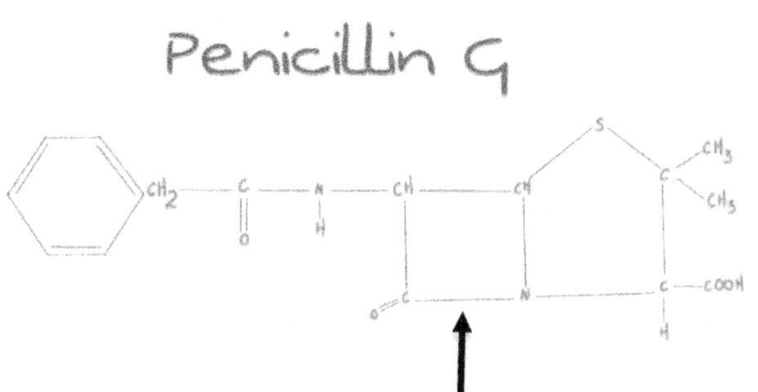

Fig. 5 Chemical structure of penicillin

However, this ring is susceptible to being broken at a point shown by the arrow. If it is broken, the penicillin is no longer able to block the formation of the bacterial cell wall. The enzyme that can break the ring is called B-lactamase and can be coded for by some bacteria in their DNA. We have seen how DNA can be passed from one bacterium to another and this mechanism for antibiotic resistance shows how it can be passed from one species to another.

Biofilms

Microorganisms that produce a biofilm are more than 1,000 times more resistant to antibiotics. We have seen biofilms forming in our mouth by bacteria, but they also grow elsewhere such as on the space station, as mentioned in Chapter 7, and on most surfaces immersed in fluids. This can include the mouth, nose, ears, lungs and bladder. The biofilm gives bacteria added protection by preventing competition from other species or preventing toxic or lethal chemicals, including antibiotics, getting too close. Biofilms can remove harmful waste products and make the uptake of nutrients easier.

Why is antibiotic resistance a problem?
Antibiotic resistance is becoming a big problem. Increasingly, some bacteria are no longer killed by certain antibiotics. In 2016, reports from America and China found bacteria that are resistant to all antibiotics. We saw in Chapter 6 how HIV developed a resistance to some of the treatment drugs. Fig. 6 shows when different antibiotics were discovered and when resistance was first reported.

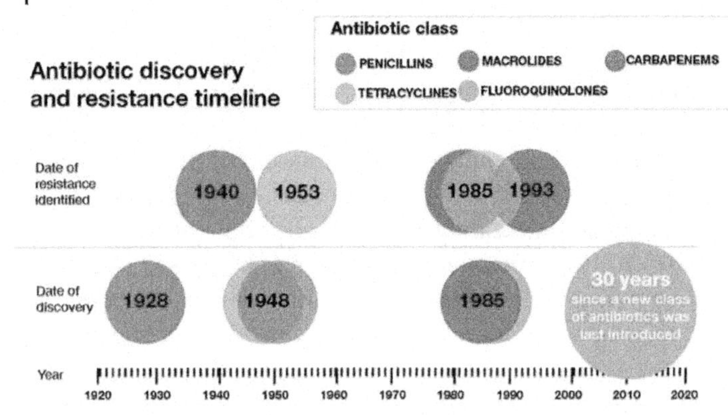

Fig. 6 Antibiotic resistance timeline (from Nature.com)

Antibiotic resistance will lead to an increase in the number of untreatable bacterial infections. Rather than the death rate from infections in the Western world remaining low, they will rise.

What are the alternatives to antibiotics?
First, we should try to prevent the infection in the first place. This means reducing our exposure to the bacteria; for example, by washing hands after going to the toilet and before preparing meals. There are vaccinations against some bacteria, which means the infection can't occur in the first place (more on vaccinations shortly). We have seen the importance of biofilms in antibiotic

resistance. Medical procedures such as inserting a catheter into the bladder create a site where a biofilm can develop.

If we do develop an infection, antibiotics are useful for killing bacteria only if they are the cause. They do not work against viral infection and are likely to remove some of the bacteria found naturally in our bowels. If antibiotics are needed, it is important that the course is completed. We have seen the difference between Gram-positive bacteria and Gram-negative bacteria and how penicillin, for example, only works on Gram-positive ones, so choosing the right antibiotic is vital.

The human body has a highly organised immune system that, if given the chance, can remove the vast majority of infections.

Some researchers are examining the possibility of using probiotics to improve the diversity and balance of the bacteria in the bowels to reduce the likelihood of infections occurring. This is being considered for cattle and poultry as well as humans.

By examining different environments, it is hoped that new classes of antibiotics will be found. Honey has long been thought to have antibacterial properties; a compound made from hops is also being examined. Polyphenols found in green tea have been used to treat the hospital superbug *Clostridium difficile*. Other scientists are examining marine sponges in Wales. One study is examining the different bacteria in surfers compared to non-surfers. It is expected that, as surfers swallow seawater, different bacteria colonise their bowels. The study is examining whether this has an effect on antibiotic-resistant bacteria.

Vaccinations

Vaccinations are a way to protect against an illness and reduce the likelihood of needing treatments such as antibiotics. We have seen how vaccinations have eradicated smallpox altogether, but how do they work?

A vaccine is a biological preparation that triggers the body into producing antibodies to protect against a specific disease. A different vaccine is needed for each disease. When the immune system has recorded and memorised the threat, it can respond more quickly if it meets the offending microorganism again. Without a vaccine, the body will make antibodies if it meets the bacteria. However, as we have seen, some illnesses kill before the body can respond. On its own, smallpox killed 500 million people in the last century before widespread vaccinations. Other infections, such as polio or diphtheria, which historically caused lots of ill health and deaths, are almost unheard of, largely due to vaccinations.

Different vaccines work in different ways:

Inactivated vaccines are microorganisms killed by heat, chemicals or antibiotics prior to use. The illnesses covered include influenza, polio and rabies.

Attenuated vaccines contain live but altered or attenuated microorganisms. This includes the measles vaccine, which was introduced in America in 1958. Prior to the introduction there were 763,094 cases and 552 deaths. Subsequently, there were fewer than 150 cases per year. Other illnesses covered include tuberculosis (see above), mumps and rubella.

Toxoid vaccines are inactivated toxic compounds that target the cause of the illness rather than the causative bacteria such as tetanus (see Chapter 11). The tetanus toxoid is a cell-free purified toxin that is combined with aluminium to improve its ability to stimulate antibody production.

Subunit vaccines are small fragments of the microorganism and include the hepatitis B virus vaccine. These can be the surface proteins from the cell wall (remember the antigens or address labels on the box in the first chapter?).

Sometimes immunisations can fail if the body doesn't respond or only has a partial response; if the strain of microorganism changes such as in HIV (in effect the 'target' no longer exists); or if there are other illnesses such as diabetes or steroid use. There is likely to be a genetic component, which means some people do not respond as well as others.

Vaccines can have some side effects. These are usually related to the body's natural response while making antibodies and can include a mild fever and headache. Some vaccines are made by using eggs to grow enough viral or bacterial protein and can affect people who have an egg allergy. Whenever a needle is put through the skin there is the risk of local swelling and bruising. Despite a controversial research paper containing falsified data from 1998 (which has since been withdrawn and thoroughly discredited), vaccines do NOT cause autism.

Herd immunity is very important for those who can't be vaccinated. These include people with illnesses such as leukaemia; patients undergoing chemotherapy; very young or very old people; and those on certain medications. Some vaccines should not be given to pregnant women.

Herd immunity 'game'

In Fig. 7, each dot represents one person. The aim is to spread a disease to as many people as possible. To do this, choose one 'person' or dot to be the first victim infected or 'case zero'. They cough and sneeze over the two closest 'people', thus infecting them and changing their colour from clear to black, as shown in the example. These two can now infect two more, and so on. A more virulent disease would affect three or four or five people rather than just two.

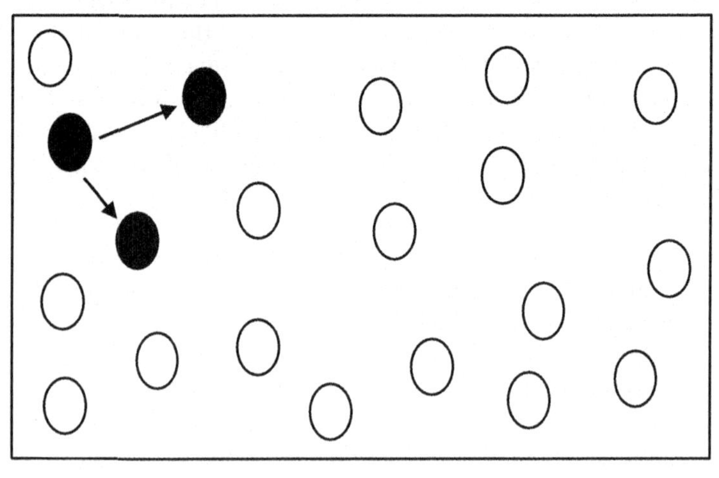

Fig. 7 Example of infected 'case zero' infecting two other people

In the first example there is no one vaccinated.

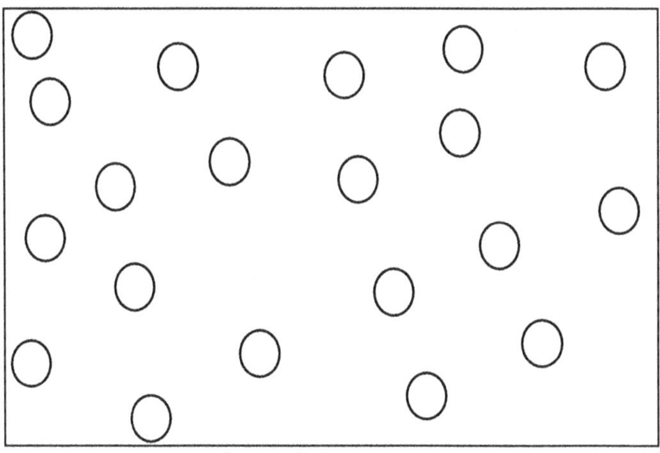

Fig. 8 Unvaccinated population

In the next example, we have a 50% vaccination rate. Those vaccinated cannot be infected and are grey. The infection spreads from case 1 to the nearest two unvaccinated or uninfected people.

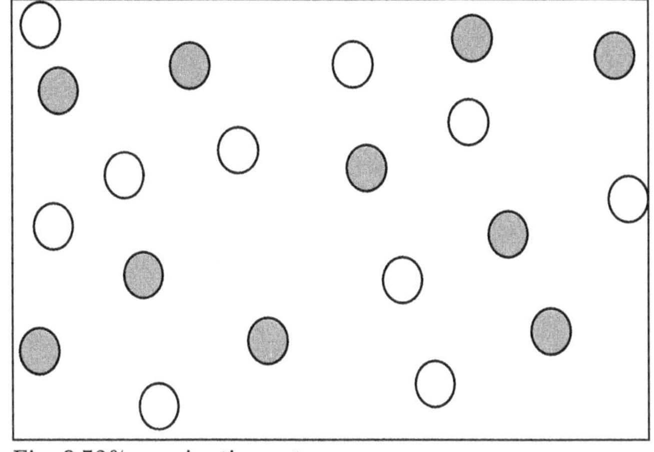

Fig. 9 50% vaccination rate

Now for a 75% immunisation rate:

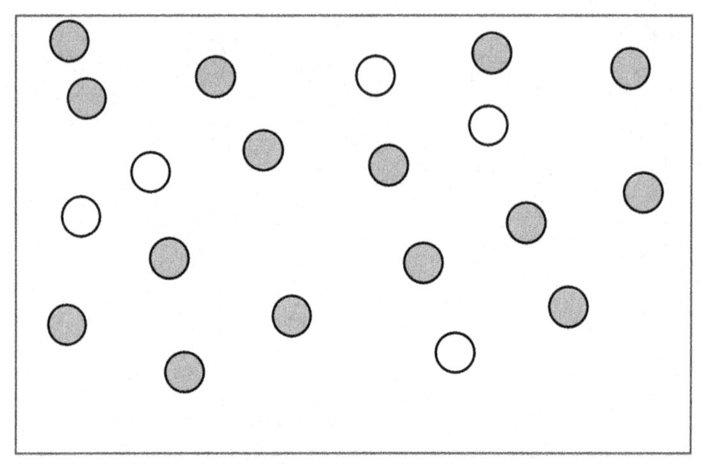

Fig. 10. 75% immunised

And finally, a 90% immunisation rate:

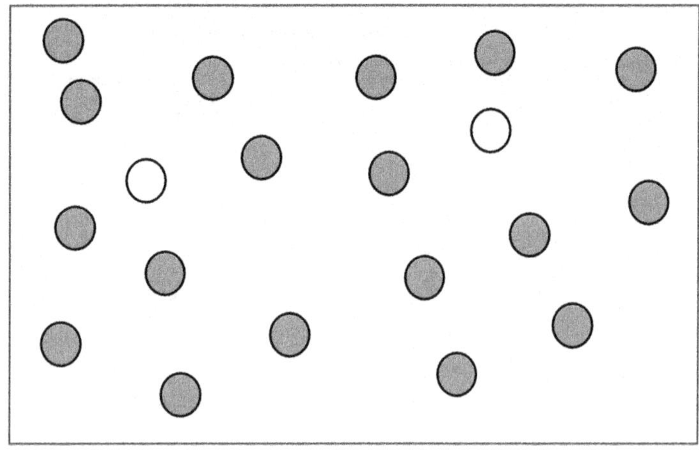

Fig. 11 90% immunised

As more people are immunised, it is more difficult for the infection to spread. At 90% vaccination, it is almost impossible to have an outbreak.

In Japan in 1974, vaccination against whooping cough was successful and reduced the number of cases to fewer than 400. Unfortunately, people then felt the vaccination was no longer needed, and within two years the vaccination rate dropped from 80% to 10%. An outbreak affecting 13,000 children and causing 41 deaths occurred in 1979. If vaccination rates drop, infection rates increase.

Summary

There are millions and billions of bacteria on and in our bodies. We usually have a healthy balance between those protecting us and those doing us harm. There is much interest in tinkering with this balance in areas around obesity, depression and other health issues. Protecting ourselves against illness is important, and vaccination forms a cornerstone of protection. Not vaccinating puts both the person themselves at risk but also allows the disease to spread to those who cannot be immunised for genuine medical reasons. Antibiotics, although useful to treat bacterial infection, can have short-term side effects on our body's natural balance and in the longer term are helping to select only the strongest bacteria to evolve and thrive.

Chapter 9 - Archaea

These things aren't even bacteria.

- Carl Woese

In 1977, Carl Woese, the American microbiologist we met in Chapter 7, made an exciting discovery that turned the scientific world upside down. He found not only a new organism, but an entire classification of organisms. They were initially coined archaebacteria, but eventually became better known as archaea - from *archaios* ('ancient' in Greek).

Although some archaea look like bacteria under the microscope, they have more in common with humans than with bacteria. Crazy though it sounds, the three groups - bacteria, archaea and eukaryotes - are totally different from each other. In 1990, analysis of RNA strands such as the 16S rRNA and 18S rRNA found archaea to be different enough from bacteria to justify a separate domain. By 2003, more genome analysis confirmed that archaea are really very different from bacteria.

Naming archaea is tricky because, due to their size and the expectations that they were just another type of bacteria, many have the suffix '-bacteria' in their name. An example is *Halobacterium*, which we will meet later.

Despite making up an estimated 20% of the Earth's biomass - predominantly as picoplankton in the oceans and ammonia-oxidising soil organisms - very little is known about archaea. As we will see, they often live in environments inhospitable to other organisms and as a result are very difficult to find and culture in the laboratory.

The very earliest cells appeared between 3.4 and 4.1 billion years ago and, although fossilised remains of early cells have been found, it is impossible to tell if they were archaea or bacteria by looks alone.

How are archaea different from bacteria and eukaryotes?

There are some similarities between bacteria and archaea such as size, speed and style of replication and some of the environments they live in, but there are many differences too. Neither bacteria nor archaea have a nucleus, organelles such as mitochondria or chloroplasts, or folded internal membranes.

Cell walls

We have previously seen that all cells have a phospholipid bilayer. This is true for bacteria and eukaryotes, but the chemical chains used by archaea are slightly different. Some have a lipid monolayer where both ends of the fatty chain are hydrophilic and can hide the hydrophobic middle part of the chain. One major factor is that the bonds within the membrane are different, with ester bonds in bacteria and ether bonds in archaea, as in Fig. 1.

Ester vs Ether linkages

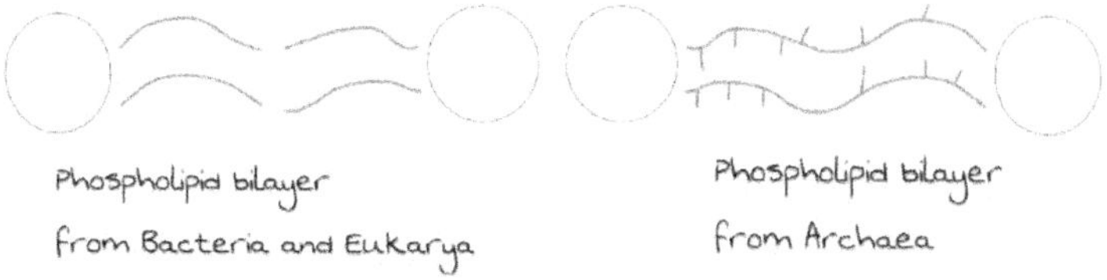

Phospholipid bilayer
from Bacteria and Eukarya

Phospholipid bilayer
from Archaea

Fig. 1 Differences in cell wall types

The types of bonds in the cells of the different domains create very different characteristics; some are listed in Fig. 2.

Archaea have different bonds in the lipid membrane, which means their lipids lack the fatty acids found in bacteria and eukaryotes.

Unlike bacteria, archaea cell walls do not contain peptidoglycan. This has implications regarding antibiotics. The antibiotics such as penicillin that target the peptidoglycan layer would be totally ineffective against archaea.

Ester bonds	Ether bonds
Found in bacteria and eukaryotes	Found in archaea
Carboxylic acid derivatives	Produced from alcohols
Carbonyl group adjacent to the -O- oxygen	Ether doesn't have the -O- functional group
Easily hydrolysed to produce an alcohol and a carboxylic acid	Ether isn't
Has a carbon-carbonyl-oxygen carbon bond and the functional group is written as RCOOR. Carbon has a double bond to one of the oxygens and a single bond to the other	Has a carbon-oxygen-carbon bond and the functional group is written as ROR
Are polar compounds	More soluble in water because water molecules can hydrogen bond with them Don't have the capability to form strong hydrogen bonds *to each other* due to the lack of hydrogen but can bond with water molecules
Needs two carbon atoms and two oxygen atoms	Needs two carbon atoms but only one oxygen atom

Fig. 2 Differences between cell wall of archaea and bacteria/ eukaryotes

There are pseudopeptidoglycan compounds found in some archaea cell walls and some also have a lipid monolayer - especially the extreme thermophiles. The differences in cell wall type also leads to an interesting array of shapes, including flat, square, box-shaped *Haloquadra,* which lives in salty pools; rod-shaped *Methanobacterium thermoautotrophicum*; spherical *Methanothermus fervidus*; irregularly shaped *Sulfolobus*; or *Methanosarcina rumen,* which are green-coloured cells with red walls.

So the cell walls are different, but what else is different between bacteria and archaea?

Ribosomes

We saw earlier that a bacterium has a smaller ribosome than a eukaryote. The prokaryotic bacteria have 70S ribosomes, made up of a large (50S) and a small (30S) subunit. Their small subunit has the 16S RNA subunit, consisting of 1,540 nucleotides bound to 21 proteins. Eukaryotes have 80S ribosomes, each consisting of a large (60S) and small (40S) subunit. Their 40S subunit has an 18S RNA, which has 1,900 nucleotides and 33 proteins.

When archaea are examined, their ribosomes have a size and shape similar to those of their bacterial counterparts - that is, a 50S and 30S - and can contain a 16S rRNA variant. However, when the sequences of RNA and protein structures are examined, they are closer to those of eukaryotes.

Archaea have evolved to work in some very extreme conditions and parts of their ribosomes are highly resistant to damage in such environments. The differences in the protein concentrations and overall rigidity of the structure makes the overall ribosome suffer fewer shape changes that could impair or stop it from functioning.

DNA polymerases

We saw in Chapter 2 that RNA polymerases were used in the replication of DNA in the cell. From what we have learned already, it is not a surprise that RNA polymerases in bacteria and archaea differ from one another. The RNA polymerase is responsible for creating messenger RNA that is then translated into proteins by the ribosome.

Bacterial RNA polymerase is simple and has only five different proteins. Some archaea, such as the methanogens and halophiles we will meet shortly, contain eight proteins. The hyperthermophilic archaea have even bigger and more complex RNA polymerases, with ten proteins. In Chapter 8, we saw that the antibiotic rifampicin works against TB: it inhibits bacterial RNA polymerase, but does not work against archaeal RNA polymerases.

In eukaryotes, RNA polymerase II is used for most mRNA transcription and it has 12 proteins. It is more similar to the RNA polymerase in the archaea than bacterial ones.

Metabolic differences

We saw that eukaryotes generated and stored energy as ATP from either respiration or photosynthesis. Some of this involves the glycolysis pathway to break down glucose such as the preferred food source of *E. coli* in Chapter 3.

Archaea use very metabolically diverse pathways to generate energy so it is difficult to generalise. Archaea do not use the glycolysis pathway itself. In Chapter 4 we saw the results from the Krebs cycle were NAD, protons and electrons being generated for the respiratory chain complex. Archaea tend not to have a functional Krebs cycle, although a few can do. There are other modified energy-generating cycles, such as the Calvin cycle or other metabolic pathways. Although some archaea use sunlight, none have so far been discovered that use classical photosynthesis.

Reproduction

Neither bacteria nor archaea have a nucleus and both expand their numbers by binary fission: these are similarities between the two domains. The DNA replicates prior to the cell being pulled apart to form two daughter cells, each with a copy of the DNA inside.

Movement

Both some archaea and some bacteria have a tail that propels them through their environment. Although virtually identical to each other in looks, they appear to have descended from different evolutionary pathways. Archaea use theirs to move *towards* the extreme environment in a process known as thermotaxis.

Subdividing archaea

We have seen previously that bacteria were originally dividing into groups based on whether or not they had a cell wall that was Gram-positive or Gram-negative. Archaea are not so easily divided on looks alone. Using genetic analysis, Fig. 3 shows a representative tree.

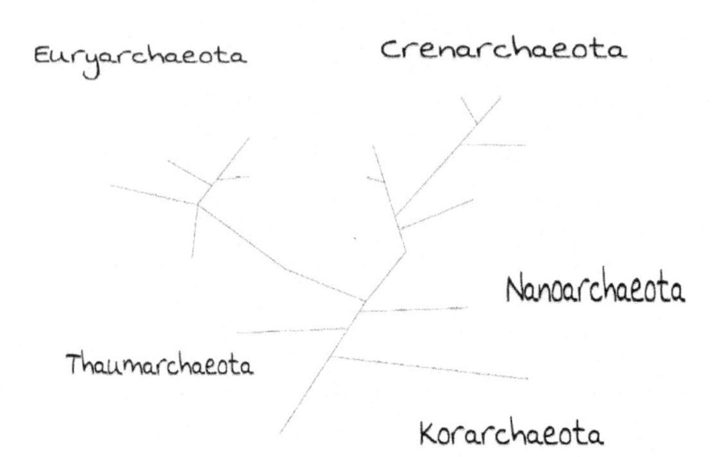

Fig. 3 Tree of archaea

The domain Archaea is divided into (at least) two main kingdoms: Crenarchaeota and Euryarchaeota.

There are likely to be at least three further groups, but debate continues as to whether they are going to be separate kingdoms or phyla. They are currently named as Thaumarchaeota, Nanoarchaeota and Korarchaeota. The latter two groups have yet to be grown in pure laboratory cultures.

Despite their abundance, no member of the Archaea domain has been known to be part of a food web, although some methane-consuming archaea can be consumed by marine worms found near Costa Rica. It is thought that this is an essential way that marine methane is prevented from entering the atmosphere.

Euryarchaeota

These occupy many different niches and have a wide variety of metabolic pathways. They can be divided into methanogens, halophiles, sulphate reducers, and many of the more extreme thermophiles.

Methanogens

Methanogens are archaea that are methane (CH_4) generators; they produce the gas as a by-product of their energy metabolism. They occupy many different habitats including marshes and swamps, where they produce 'swamp gases', animal digestive tracts, the hydrothermal vents we will see in Chapter 12, and human waste-management facilities. They often live in a symbiotic relationship with fermenting bacteria. Up to 600 litres of methane can be produced every day from a cow's stomach. Methane is a more damaging gas than carbon dioxide when it comes to global warming, and increasing the global cattle population will result in higher methane production. *Methanopyrus kandleri* was discovered in the black smoker of the Kairei hydrothermal field in the Indian Ocean and can survive at 122°C - the highest recorded temperature for any organism.

Halophiles

Extreme halophiles are salt lovers and live in habitats that are extremely saline. In Chapter 12 we will see the definitions of freshwater (<0.05%), brackish water (>3%), saline (3-5%) and brine (>5%) salt concentrations. Extreme halophiles require at least 9% salt for growth and some may tolerate near-saturation at about 32% salt! Places for growth include the Dead Sea and Great Salt Lake, but they can also be found in heavily salted foods. An example includes *Halobacterium*; this contains a purple pigment, bacteriorhodopsin, which provides it with chemical energy by pumping protons outside of the cell and generating ATP as they return. This pigment is similar to the one found in our eyes called rhodopsin.

These archaea make their energy from light but not by photosynthesis.

Sulphate reducers

Sulphate reducing archaea were first identified from hydrothermal vents by 16S-rRNA sequencing in 1987. The genus was labelled *Archaeoglobus.* They were found in high temperature environments and preferred 82°C. Their ideal pH was around 6 wth 1.5-1.8% salinity

Extreme thermophiles

Extremophilic archaea are even more extreme than their bacterial counterparts. Thermoplasmatas thrive at extreme temperatures and/or acidities. For example, *Thermoplasma* lives in piles left over from coal mines that have a high temperature and are highly acidic. They do not have a cell wall but instead have a unique membrane made of a lipoglycan - a fat and sugar molecule joined together. This is likely to allow stability at such high temperatures. There are two species, *T. acidophilum* and *T. volcanium;* they live in oxygen-free environments and use sulphur and carbon for energy.

Picrophilus is an extreme acidophile. It prefers to grow at pH 0.7 and reportedly can grow at pH0.06. It was first found in Japan in hot acidic springs and is related to *Thermoplasma. Pyrococcus furiosus* can double in only 37 minutes and has enzymes that make use of tungsten, which is very rarely found in biological organisms. Their enzymes can be used in the PCR reactions we meet in Chapter 14.

Crenarchaeota

These are thought to be similar to the ancestors of archaea. The species so far discovered are thermophiles or extreme thermophiles. *Pyrolobus fumarii* survive at

113°C, but have spent ten hours in an autoclave at 121°C and survived and continued to multiply. Originally thought to use sulphur for energy, some more recently discovered species have been identified in the oceans and are expected to be the most abundant form of life below 300m. Some of these, such as *Nitrosopumilus maritimus*, can oxidise ammonia for energy at a 'chilly' 28°C. The majority are strict anaerobes and have also been found in deep-sea volcanoes and in Yellowstone.

There are two subdivisions, including *Sulfolobus* and *Thermoproteus*. The former was initially found in Italian sulphuric springs at 80°C and pH2-4, but have subsequently been found around the world. They can gain energy from organic compounds such as sugars as well as inorganic sulphur or iron compounds. *Sulfolobus acidocaldarius* can leach copper and iron from ore.

The smaller groupings include the following:

Korarcheota

These were discovered in 1996, in a hot spring in an obsidian pool at Yellowstone. They were discovered by using environmental sampling of their ribosomal DNA. At least 19 different groups have been discovered and depend on geography, temperature and salinity levels. There are some features that fit with both Crenarcheaota and Euryarchaeota, which suggests a very early evolutionary branch.

Thaumarchaeota

These are hyperthermophiles that have a different genetic sequence to Crenarcheaota and may be important in nitrogen and carbon cycling. Genetic analysis found an enzyme called a topoisomerase, which helps to wind and unwind DNA; this had only previously been found in eukaryotes.

Nanoarchaeota

This is a very lonely classification as there is only one member and it is the smallest archaea known: *Nanoarchaeum equitans*. It was found in a hydrothermal vent off Iceland in waters with a pH6, salinity 2% and a temperature of 80°C. It has only 540 genes in a single circular chromosome (490,885 nucleotides), compared to 4,485 genes in *E. coli* bacteria and 23,000 in humans. The next smallest is the bacteria *Mycoplasma genitalium*, which has fewer genes (475), but more base pairs (580,070). Remarkably and almost unbelievably, *N. equitans* does not have a 16S rRNA gene. This expected sequence is undetectable by current methods. With no rRNA gene there can't be a ribosome, and no ribosome means no proteins. In earlier chapters we saw that these are vital for a cell to survive. So how does it manage? It seems that about 95% of its gene codes for enzymes to repair DNA and for replication. *N. equitans* is a parasitic microorganism and lives on a bacterium, *Ignicoccus,* and needs cell to cell contact. It doesn't appear to make its own ATP, nor does it have the enzymes to generate energy from methane or sulphur; it steals biomolecules such as nucleotides, lipids and amino acids from its host. So far, there haven't been any adverse effects noted on the *Ignicoccus.*

It is possible that, just as in endosymbiosis theory, the bacteria-cum-mitochondria lost genes to the host cell, *N. equitans* may have eliminated unnecessary genes or transferred them to *Ignicoccus*. Another option is that it was one of the first microorganisms on the planet.

It can be difficult to imagine the small sizes here. *N. equitans* is only 400nm in size or 0.4μm. By comparison, a bacterium is about 1-2μm; the diameter of a red blood cell is 8.4μm; and a grain of sand is a big 20μm. A grain of salt, at 0.1-0.3mm or 100 μm-300μm, is positively huge!

The smallest cell ever observed belongs to the kingdom Crenarchaeota. *Thermofilum* are rods of 0.17μm diameter and from 1μm long. They are small so they can pass through the pores of rocks in order to colonise deep underground environments.

How do archaea survive in hostile environments?

We have seen that there are several vital compounds for a cell to survive: DNA, RNA and proteins. We have seen previously that the RNA polymerases and ribosomes are different and that due to the structures they are more rigid and more stable as a result.

DNA stability

In Chapter 14 we will see that at temperatures above 95°C there is a breakdown of the DNA and it separates into two strands. The same does not occur in thermophilic archaea because they contain a unique 'reverse DNA gyrase' that constructs positive supercoils that stabilise the DNA. We will see that eukaryotes use histones to store DNA. Archaea use the same histones for stability and compaction of the DNA. A subtype of archaea we met above, the Crenarchaeota, also have DNA binding proteins, which bind to the DNA and raise the melting point by a further 40°C. There is a chemical within the cytoplasm that prevents damage to DNA and prevents permanent shape changes to the coils.

Protein stability
Archaea use a similar amino acid composition to construct their proteins as in bacteria, but due to the way they are folded they become more stable at higher temperatures. There are some special proteins called 'heat shock proteins' that can refold damaged and denatured proteins and perform a sort of repair function. Some amino acids are hydrophobic; proteins with these at their core have a lower risk of unfolding.

Are archaea a threat?
So far, no archaea have been found that cause damage to other organisms. This may in part be due to the locations where they thrive, where little other life ventures or survives. As more of the planet is explored, humankind can disturb or transfer the archaea to new environments. Archaea lack peptidoglycan in their cell walls (unlike bacterial walls, which have this sugar/polypeptide compound present), so are not damaged by antibiotics.

Some enzymes found within archaea can kill certain species of bacteria, such as *Staphlococcus aureus, Bacillus anthacis* and *Clostridium difficile*. This could make a difference in the hunt for new antibiotics. Anthrax is discussed in Chapter 11.

Summary
Archaea are often found in environments where other organisms do not and cannot survive. As a result, there is less competition for resources. Although they can also exist in less extreme environments, they tend to be out-competed.

Historically, it is thought that from LUCA (see Chapter 5), there was an early branch that split bacteria from archaea. It is from the Archaean branch that eukaryotes subsequently branched further and evolved. This does not mean that all archaea will share all their properties with all eukaryotes. The infolding of the membrane we saw in Chapter 1 that occurred about 1.7 billion years ago happened in the archaeal lineage.

Chapter 10 - Biodiversity

...while nature has considerable resilience, there is a limit to how far that resilience can be stretched. No one knows how close to the limit we are getting.

- Douglas Adams

Across the planet, there is an amazing range and number of organisms, from those we can see, such as whales or elephants, to those we can't, such as bacteria and archaea. We will develop an understanding of the interdependency each organism has on its neighbour and the types of relationships they can have, but first some definitions of the different parts of the ecosystem.

Biodiversity refers to the variety of organisms within a particular ecosystem. There is a breath-taking array of ecosystems on the planet. These can be divided into land-based (terrestrial) and water-based (aquatic).

Water-based ecosystems are further subdivided into freshwater and salt water (including the marine ecosystem explored in Chapter 12). Within these categories, there are further subsections. Salt water covers 71% of the planet's surface. It can be categorised into deep and shallow ocean, inter-tidal, estuaries, coral reefs, salt marshes and hydrothermal vents. Each of these ecosystems offers a unique environment for species to live in. Only 0.8% of the Earth's surface is freshwater. This includes pools, pond and lakes; streams and rivers; and wetlands or land saturated with water.

On land there are forests - evergreen, deciduous, tropical or temperate; deserts - which cover 17% of the Earth's surface; grasslands, including savannas and prairies; and mountains.

An **ecosystem** consists of all the living organisms and non-living factors in a given area. Ecosystems are not a set size; they can range from a decomposing tree, a pond, or a meadow to an entire forest or ocean.

A **habitat** is the non-living part of an ecosystem and forms the home for the organisms. It is the physical characteristic of the area such as the rock type, amount of light, or temperature.

A **community** refers to all the populations of different species living in a specified area. This could include all the creatures living on a tree or all the organisms living in a pond. Competition occurs within and between these populations for the means of survival. Within a single community, one population is affected by other populations - the biotic factors - in its environment.

A **population** is the group of individuals of the same species; for example, the number of mountain pine beetles on a pine tree, or the number of brown trout in a pond. Populations within communities are affected by, and affect, the abiotic factors in an ecosystem.

A **niche** is the part of the ecosystem that a particular organism occupies. For example, the finches we saw in Chapter 5 occupy different niches depending on beak size and type of food available.

Abiotic and biotic factors

Two main factors interact within each ecosystem: abiotic and biotic factors.

Abiotic factors are physical characteristics covering anything non-living or non-biological such as water, light, temperature, oxygen levels or acidity levels (pH). The soil type varies, ranging from sands to clays to silts, and reflects the particle size, chemical make-up and amount of biological or organic material contained. Weather and climate are important, with light, temperature, wind and humidity playing their roles, and so is the amount of nutrients available, especially carbon, nitrogen, phosphorus and sulphur.

Biotic factors refer to the living parts of the environment and the interactions and relationships between the different organisms. These include everything ranging from the biggest plants and animals to the smallest microscopic bacteria and archaea.

The nitrogen cycle: nutrient recycling

The availability of nutrients in an ecosystem is a vital abiotic factor. We have seen in the earlier chapters how some atoms such as oxygen, carbon, nitrogen, phosphorus and sulphur are vital building blocks to larger and more complex molecules. We will look at the nitrogen cycle in more detail next, while the carbon cycle is examined in Chapter 11.

Air is 78% nitrogen; it exists as two nitrogen atoms bonded together as N_2. Covalent bonds hold the atoms together and involve sharing three electrons, which make the bonds very difficult to break down and release usable nitrogen atoms. As a result, the nitrogen in the air is known as inert or unreactive. Animals, including humans, cannot use any of this nitrogen, yet nitrogen is an essential element in making amino acids and proteins.

The process of taking nitrogen from the atmosphere and turning it into a form that plants and animals can use is called nitrogen fixing. Bacteria in the soil perform this task biologically, although there is an industrial process too, as we will see shortly. On land, two types of bacteria can fix nitrogen: bacteria living freely in the soil, and bacteria that form symbiotic relationships with plant roots. In the water, cyanobacteria is a free-living nitrogen fixer. A simple nitrogen cycle is shown in Fig. 1.

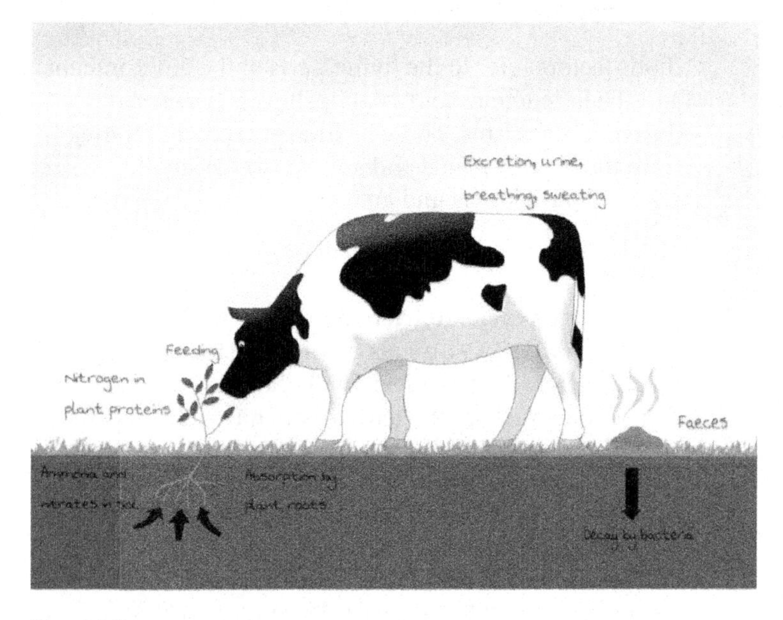

Fig. 1 Nitrogen cycle

Nitrogen is 'fixed' as bacteria convert atmospheric nitrogen into a compound that can be used by plants during growth. This compound is ammonia and is written chemically as NH_3. The plants are then eaten by primary consumers, which supply the next part of the food chain with a usable form of nitrogen.

Other bacteria can decay plants and animals and break them back down into usable nitrogen such as ammonia. Urine contains nitrogen as a compound called urea.

Nitrification occurs when these organic products are converted into another nitrogen compound - nitrates.

Denitrification is where other bacteria in the soil break down nitrates from the soil and return nitrogen to the atmosphere. A more detailed summary can be seen in Fig. 2.

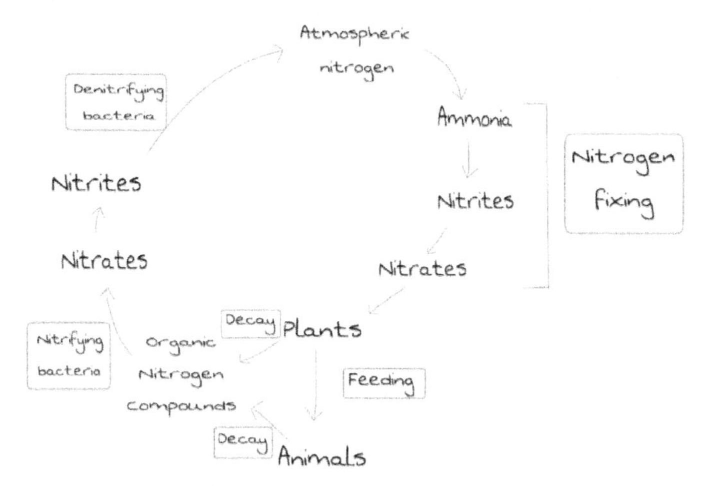

Fig. 2 More complex nitrogen cycle

Fertiliser is a way to add nitrogen to the soil in the form of ammonia, ammonium nitrate or urea. These compounds are created by factories 'fixing' nitrogen from the air. Farmers spread these usable forms of nitrogen onto their fields for their crops to grow. The industrial reaction is called the Haber process: nitrogen is combined with hydrogen at high temperatures (400–500°C) and high pressures to form ammonia. The hydrogen is usually formed from methane in natural gas. This produces carbon dioxide in the process and takes a lot of energy. Creating the high pressures and temperature needed is also an energy-intensive process. It is estimated that 3–5% of the world's gas supply and 1–2% of the world's electricity supply is used to create fertilisers. An estimated 50–80% of the nitrogen found in the human body today was originally derived from the Haber process.

More details

We can start to add the chemical formulae to the nitrogen cycle, as in Fig. 3.

Nitrogen fixation or N_2 -> NH_3 or NH_4^+

Nitrogen is fixed in one of three ways: biologically by nitrogen-fixing bacteria and archaea such as *Rhizobium;* atmospherically by lightning; or industrially in the Haber process.

About 5–8% of nitrogen is fixed by lightning. The huge temperatures and energy that are released during a lightning strike can break the covalently bonded N_2 and form nitrous oxides with atmospheric oxygen. Nitrogen can also be fixed during forest fires or in lava flows, but these contribute even smaller proportions of the global fixed nitrogen.

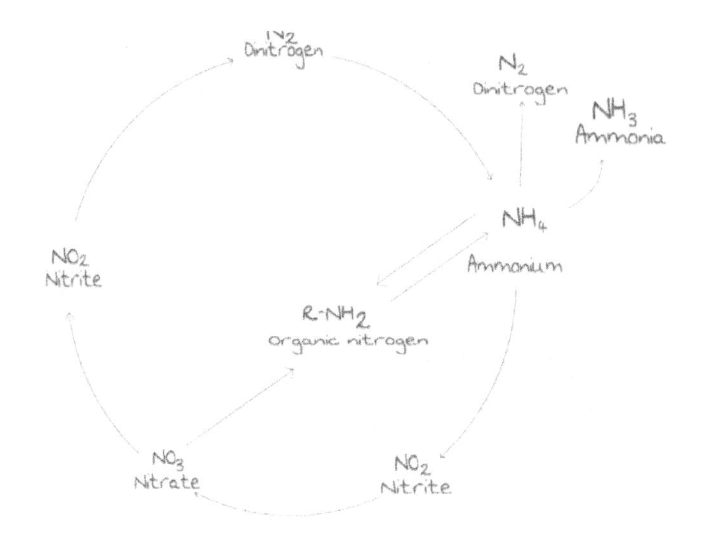

Fig. 3 Nitrogen cycle

Biologically, symbiotic relationships are formed between some bacteria and the root systems of legumes - the group of plants including beans, peas and clover. These bacteria are seen as nodules on plant roots, as in Fig. 4. The symbiotic relationship allows the plant to gain nitrogen and the bacteria to gain carbon in the form of sugars and carbohydrates and have a secure environment to live in.

Other bacteria, such as some species of *Citrobacter* or *Enterobacter*, form symbiotic relationships with animals such as termites and wood-eating ship worms. They fix nitrogen via enzymes called nitrogenases. (microbewiki.kenyon.edu)

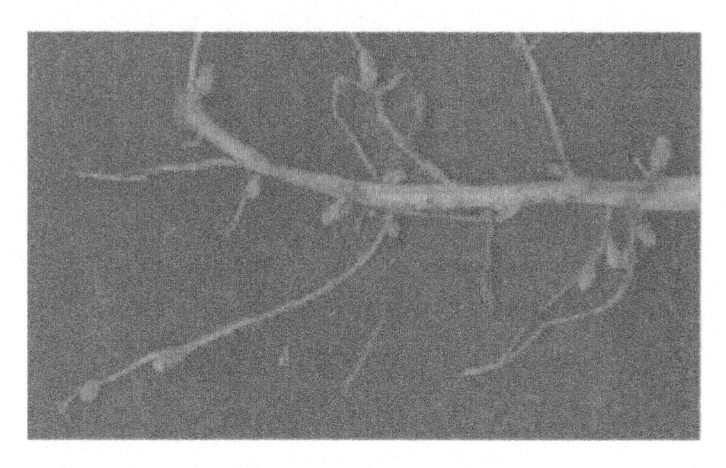

Fig. 4 Nodules on a plant root

Following fixation is **nitrogen uptake**, which is the plant using the nitrogen compound. This is represented as:

NH_4^+ -> organic nitrogen as amino acids, proteins, and DNA.

Decay or nitrogen **mineralisation**. This is the reverse of nitrogen uptake and converts organic nitrogen back into NH_4^+.

Nitrification is one of two reactions involving different bacteria: either NH_4^+-> NO_2^- or NH_4^+-> NO_3^-. The former reaction uses nitrifying bacteria such as *Nitrosomonas*, the latter by *Nitrobacters*; in each case, the bacteria gains energy from the process. As this process requires oxygen, it only occurs in oxygen-rich environments such as waterlogged soils, sediments and the surface layer of soils or rivers. Archaea *Crenachaeota* exists in soil and water; it is more abundant that the nitrifying bacteria and performs the same role.

Denitrification is the return of nitrogen back to the atmosphere as N_2. Bacteria including *Pseudomonas denitrificans* break down NO_3^- -> NO_2^- ->NO ->N_2O ->N_2. This is an anaerobic reaction and releases three gases into the environment: nitric oxide (NO), nitrous oxide (N_2O) and nitrogen (N_2). Nitric oxide is a component of smog and can combine with water to form nitric acid - one of the acids found in acid rain. It can also react with ozone (O_3) and deplete the ozone layer. Nitrous oxide is an important greenhouse gas and is 298 times more potent than carbon dioxide.

Fixed nitrogen can be lost at other stages of the nitrogen cycle. One important loss is due to leaching. This often occurs to NO_3^- but also ammonia - NH_4^+. Ideally, land used for crop growth would only have the amount of nitrogen that the plants actually need. New technology using drones to measure the amount of chlorophyll in crops is being used to let farmers know which crops need watering or need fertiliser added. This reduces the excess fertiliser being added if it isn't required by the plants. Leaching only occurs when water is passing through the soil. Elsewhere, deforestation contributes to nitrogen loss. The loss of rainforests causes the fixed nitrogen in the soil to leach out. When land loses structure and plants in the soil, dissolved nitrogen compounds wash into the local watercourses.

Water leaching into local water supplies is one cause for blooms of algae. Basically, when large quantities of nitrogen are suddenly available, this stimulates an explosion in algae growth. In turn, the watercourse is depleted of oxygen, causing other plant and cell life to die and decay. The algal blooms also block out the light needed for the other plants to grow. The death and decay releases more nitrogen due to the decomposition of the plants and animals, and this prolongs the cycle. By the time the bloom of algae has used all the nitrogen, there is little other life remaining in the river or stream.

Exponential growth and limitations

Why hasn't one species of plant or animal populated the entire Earth? In some conditions, we can see explosive growth of an organism such as above when algae has a new-found supply of nitrogen. Experiments using yeast in a flask with an endless supply of food show that the number of cells can double every few hours. Equally, in ideal conditions and left unchecked, a population of rabbits has a similar rapid reproductive rate, as seen in Fig. 5; this is called exponential growth.

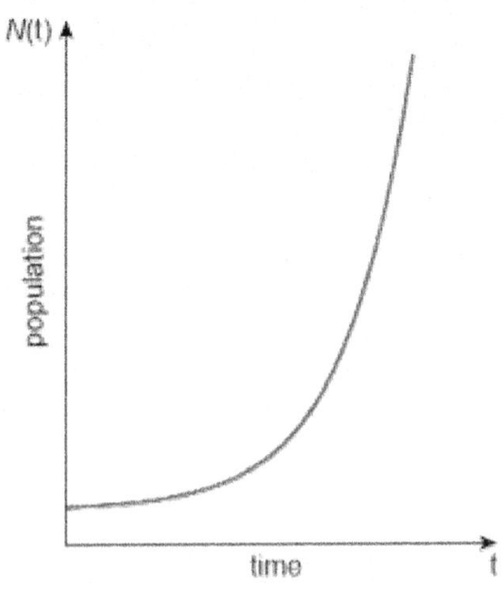

Fig. 5 Exponential growth

We can use some maths to calculate exponential growth.

If we represent the number of organisms as N, in the beginning we have N_0. If we multiply this starting number by the reproductive rate (R) of the organisms, we can work out the number of organisms in the next generation, labelled N_1,

$$N_1 = N_0 \times R$$

where N_0 = initial population size, N_1= individuals after one generation and R = reproductive rate

An example: if the initial population of an organism is ten, and each organism produced four offspring in its lifetime before dying, then the number of organisms after one generation would be:

$$N_1 = 10 \times 4 = 40$$

In the second generation, the calculation would be:

$$N_2 = 40 \times 4 = 160$$

However, in reality, exponential growth doesn't occur. Most populations reproduce until they reach the limitations imposed by their environment such as running out of resources such as food, light, oxygen, water or space. At this point, a logistic growth model takes these factors into account. As the population increases, growth rate slows and eventually the population size plateaus.

As reproductive rates gradually slow, the shape of the graph changes to more of an S-shaped curve (also known as a logistic curve), as seen in Fig. 6.

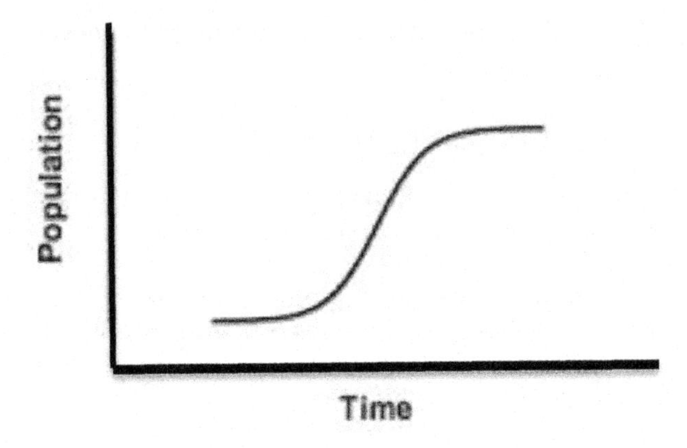

Fig. 6 Logistic growth

The upper limit is known as the carrying capacity of the environment. In the case of yeast in a sugary solution, the limitations can be: using up all the food; overcrowding; or the production of toxic and polluting waste products. With yeast, the waste product is alcohol - a fact that humans have been exploiting for several millennia!

With rabbits in a field, the limiting factors may be food supplies, overcrowding, or something harmful such as a predator or disease.

Interacting organisms

All life begins with organisms that can create energy from abiotic factors. These are known as primary producers, but can also be called autotrophs ('auto' = self; 'troph' = feeder). We are most aware of plants as producers as they use chlorophyll and photosynthesis to create carbon-based nutrients from sunlight – which we know as food! Other organisms – especially the extremophiles we saw in Chapter 7 - can use other chemicals such as methane to drive energy production and hence life. We saw that some of these microorganisms grow in places with no light, such as caves or deep ocean beds. In marine environments, it is mainly algae or phytoplankton that are primary producers.

After autotrophs come heterotrophs ('hetero' = other). These are consumers and digest the producers for energy. First-order consumers are those that eat the producers, such as rabbits eating grass. Second-order consumers are animals that prey on the first-order consumers, such as foxes eating rabbits. The top consumers are those that have no natural predators, such as wolves, lions or humans.

When the producers and consumers are connected together, they form a food web sometimes represented as a food pyramid. A pyramid does not reflect the extreme complexity when all organisms of an ecosystem are considered. Fig. 7 shows a simple food web.

Nutrients are recycled back from the producers and consumers back into the soil; this is often done by fungi and bacteria. These break down and decompose the complex organic molecules that have been formed. The organisms responsible are called reducers, decomposers or saprotrophs ('sapro' = rotten).

A complete cycle for nutrients is created, from producers to consumers to decomposers and back to producers again. If parts of the food web are removed or damaged, the whole cycle is affected, as we will see.

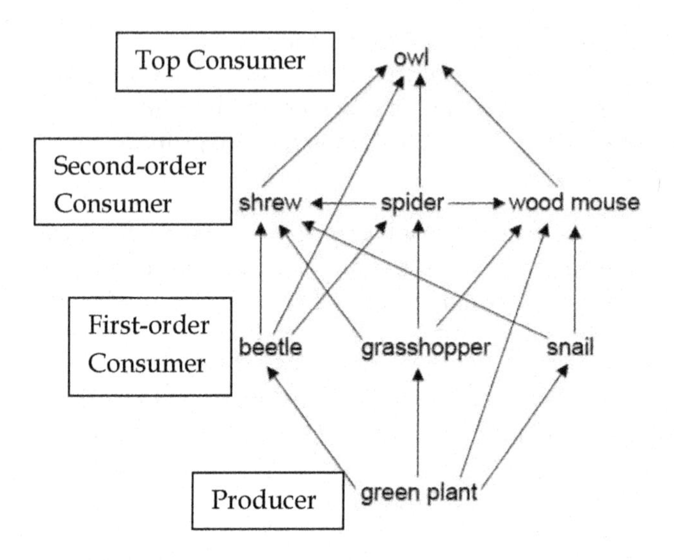

Fig. 7 Food web

Measuring diversity

To understand the ecosystem we need a way to calculate the different species present and the balance between them. Diversity refers to the variety of organisms present, although there are different types of diversity. Ecological diversity covers different habitats, niches and species interactions. Species diversity refers to the different types of organisms and the relationship between them. Genetic diversity concerns the different genes and combinations of genes within populations of organisms.

How much diversity is there?

The entire Earth is covered in habitats, ranging from the deepest trenches of the oceans to the tops of mountains, and from the hot deserts to the cold poles. Life exists in all of these places. Although each habitat has a unique population of species, most life is concentrated in warmer climates. Enzymes have optimum operating conditions; this plays an important role in finding the optimum conditions for each species.

Each year some new species are found but other species become extinct. The number of known species is about two million, but the number yet to be discovered is an estimated 100 million.

Many of the unknown species are expected to be single-celled bacteria and archaea. When discussing extremophilic bacteria, precisely because they live in extreme environments, they are difficult to find and difficult to grow. We have seen how the 16S rRNA gene can be used to identify organisms in an environment even when we can't find the organism itself. If cells from the bottom of the ocean were brought to the surface, the change in pressure would be likely to make them burst open. The organisms would no longer be present, but their DNA would survive. Genetic analysis techniques are useful to help estimate the total amount of biodiversity.

Why is biodiversity important?

Ecosystems with many different species are more resilient to events than single-species systems or monocultures. For example, a wildflower meadow may have 30 different flowering plants and 100 different grasses. To this number, the hundreds of insects and primary consumers can be added, then the birds and bats; suddenly, a meadow can support and maintain thousands of different species. Contrast this to a field of wheat. The field may have been sprayed with insecticide or pesticide to remove any insects prior to sowing the crop, as farmers are concerned they may damage the financially valuable crop. The single plant species of wheat is grown with the aim of producing as high a yield as possible. Further insecticides and fertilisers are added to the field to optimise growth.

Biodiversity is vital for maintaining balance between the organisms in the community. Once a species has been removed from an ecosystem it cannot usually return. The ecosystem can only use the resources and organisms that survive. Trying to convert a wheat field back into a wildflower meadow is very difficult and can take many decades.

A diverse ecosystem is more likely to be able to resist invasive species. They are more stable and can be more productive. The greater the biodiversity of an ecosystem, the more likely it is to recover from stresses such as adverse weather or diseases.

Reducing biodiversity is shown in ecosystems other than a farmer's field. In Indonesia there has been a decimation of natural rainforests and the land used for the planting of crops particularly to produce palm oil. Palm oil is used in a huge variety of household products and foods. Indonesia contains 18,000 islands and until the 1960s had forest covering about 80% of it. Globally, 12% of all mammal species, 16% of reptile species, 17% of bird species, 24% of fungi and 33% of insect species were found on these islands. Since then, nearly 50% of the rainforests have been cleared, with the trees being felled for the making of paper and plywood and the space used for planting palm oil trees. Cleansing the land and replacing the natural diversity with a single monocrop will have effects felt for many generations to come. For example, when the land is cleared to the extent that the orangutan is made extinct in Indonesia, it will not be able to repopulate the region naturally. The same is true for many smaller creatures, plants and organisms.

Loss of biodiversity also happens at sea. For example, in recent years coral bleaching has taken place on the Great Barrier Reef off the coast of Australia (see Chapter 12). This is when the algae (zooxanthellae) that live on the coral in a symbiotic relationship are expelled and can occur due to changes in light, temperature or nutrient levels. Removing one part of the food web impacts on the diversity of the remaining organisms. Bleaching primarily occurs due to the rise in sea temperatures, but can also take place when the sea cools.

Ash dieback

Ash trees in the UK have been affected by a disease called 'ash dieback' or Chalara dieback of ash. A fungal disease, first noted in the UK in 2012 in Buckinghamshire, had been causing problems in Europe since the early 1990s. It is most devastating to young ash trees. It seems to weaken the ash tree's ability to protect itself against pathogens or pests. It is spread mainly by the wind dispersing spores but also by the movement of infected logs or ash plants. If a woodland consisted of a monoculture of ash trees, the effect of this fungus would be to wipe out all the trees. A diverse mix of trees means that although the ash would be affected, the other trees would survive. These surviving trees would continue to provide a habitat for insects and other plants and animals.

There is some genetic diversity within the ash population and it is highly likely that a small fraction have a natural immunity or resistance to the fungus. These survivors will be able to reproduce and their offspring will inherit the genes for immunity, meaning that the population is likely to recover over time.

More details

Chalara is caused by the fungus *Hymenoscyphus fraxineus* (previously called *Chalara fraxinea*). It produces spores as part of its lifecycle that are dispersed by the wind. The two parts of the lifecycle are the primary phase, or infection of the ash tree, which affects the bark; the second phase involves spores being produced by reproduction on the fallen leaves from the previous year and produces fruit bodies. Spores are produced by these fruit bodies between June and September and a tree needs a high dose of spores to become infected. Once infected there is no cure. If a second infection of honey fungus (*Armillaria*) is present, the ash tree succumbs and dies far quicker.

Another devastating illness affecting a species of tree is Dutch elm disease, as discussed in Chapter 13.

Calculating biodiversity

If we consider a children's ball pit with many different coloured balls inside, then just by observing we can tell whether there is more of one colour or another. If we start with a single colour or monoculture, as seen in Fig. 3, we

can see that this is 100% white - there is no diversity of colour. Each of these balls could represent a wheat plant in a farmer's field or a palm tree in the plantations of Indonesia.

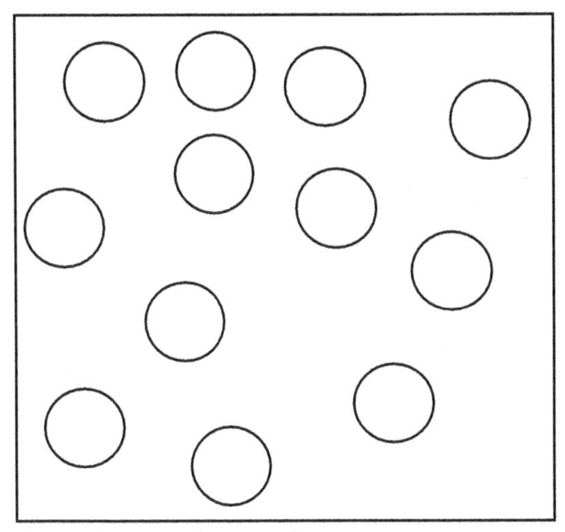
Fig. 3. Monoculture – a)

If we take two more ball pits, this time with balls of four different colours, we can see whether they are the same or different, as in Fig. 4. There are 12 balls in total, but each group has different numbers of individual colours.

The population b) is dominated by dark grey balls. There are only four out of 12 balls that aren't dark grey. In the second population, there is an equal number of different colours used; this is an even proportion.

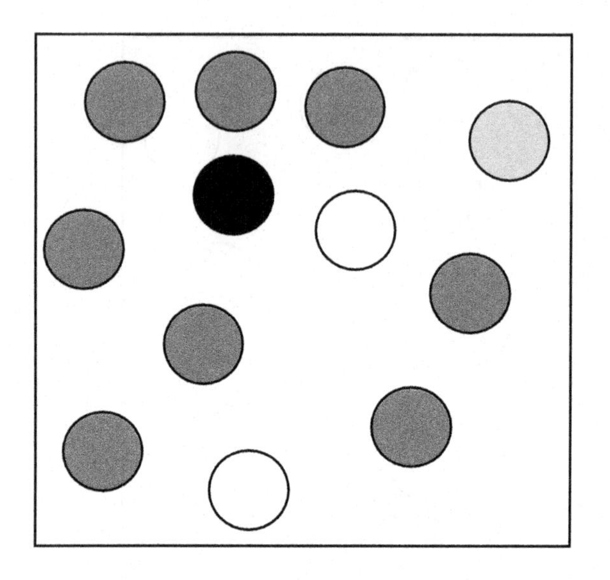

Fig 4. Two different population b) above and c) below

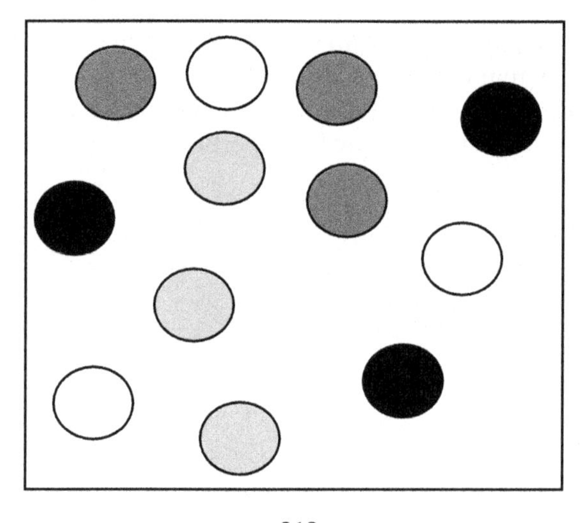

If instead of balls we use types of tree in a rainforest or woodland, variety of fish on a coral reef or bacteria in our bowels, then we can start to see the complexity of species diversity around us.

How do we calculate diversity?

When it is a simple 'four colours in a ball pit' we have a gut instinct and can see visually which is most diverse. However, it is more difficult if we return to the wildflower meadow with hundreds of different species of flower, grass and animal. Some simple maths can help work out which population has the most diversity or the most evenness.

First of all, we can calculate species richness (the number of species present), and label it 'S'. This is easy for the monoculture as there is only one colour, so S=1. For our next two populations, there are four different colours or 'species', so S=4.

Next we list the different colours and count how many of each colour are present. This can be put into a table such as Table 1.

Table 1

	A	B	C
SPECIES RICHNESS (S)	1	4	4
WHITE	12	1	3
DARK GREY	-	8	3
LIGHT GREY	-	2	3
BLACK	-	1	3
TOTAL BALLS	12	12	12

To work out the proportion of each colour of the whole population, we take the number of individuals and divide by the total. We label proportion as pi.

For the monoculture a), it is 12/12 = 1 as it is 100% white.

For population c) as there is the same number of each species i.e. 3, the sum is 3/12 = 0.25. We can tell that one-quarter of the balls are grey or black.

For population b) it is slightly trickier, but can be put into a table such as Table 2.

Table 2

COLOUR	NUMBER OF BALLS	TOTAL BALLS	EQUATION	RESULT (PI)
WHITE	2	12	2/12	0.16
DARK GREY	8	12	8/12	0.66
LIGHT GREY	1	12	1/12	0.08
BLACK	1	12	1/12	0.08

This shows that two-thirds or 0.66 of population b) are dark grey.

Simpson index

In 1949, Edward Simpson used these numbers and a simple calculation to measure how concentrated one species is in a population.

His equation became known as the Simpson index or D (for Diversity) and is written:

$$D = \sum pi^2$$

The Σ means 'to sum' or to add up.

We have worked out our proportion or pi values above. If we square these and add them together, we have calculated the Simpson index.

For population a) D is the sum of 1^2, which is 1.

For population b), D is the sum of $(0.08)^2 + (0.66)^2 + (0.08)^2 + (0.16)^2 = 0.474$.

For population c), D is the sum of $(0.25)^2 + (0.25)^2 + (0.25)^2 + (0.25)^2 = 0.25$.

Counterintuitively, the lower the number, the higher the diversity. This feels confusing but shows that population c) is the most diverse.

As a lower number to represent a higher level of diversity doesn't feel right, more commonly the 1-D amount is used. This is the Gini-Simpson index and gives a result that feels easier to use – that is, the higher the number, the more diverse the population.

For our populations, the results of the Gini-Simpson index would now read:

a) 0.

b) 0.526.

c) 0.75.

Again, this shows that population c) is the most diverse. Although we have used a small number of colours, the sums prove that our gut feelings are right. Population a) has no variety in ball colours and population c) has a better mix of colours.

These equations have, however, been superseded by several other calculations that offer a better balance between evenness and diversity and are more reflective of the populations studied. Some equations such as the Simpson index don't represent smaller, rarer species well and as a result are more commonly referred to as a **dominance index**.

Shannon-Weiner index

A better equation to measure diversity is the Shannon-Weiner index or H'. This uses the numbers we have worked out in our tables above but adds an extra calculation or two and is written as:

$$H' = -\sum (pi)(log_2 pi)$$

Table 3 adds these new calculations in:

Table 3 Shannon-Weiner index

	A	PI	LOG_2PI	$(PI)(LOG_2PI)$
SPECIES RICHNESS (S)	1			
WHITE	12	12/12 = 1	$log_2(1) = 0$	1 * 0 = 0
DARK GREY	-			
LIGHT GREY	-			
BLACK	-			
TOTAL				0

	B	PI	LOG_2PI	$(PI)(LOG_2PI)$
SPECIES RICHNESS (S)	4			
WHITE	2	2/12 = 0.16	$log_2(0.16)$ = -2.64	0.16*-2.64 = -0.42
DARK GREY	8	8/12 = 0.66	$log_2(0.66)$ = -0.60	0.66*-0.60 = -0.40
LIGHT GREY	1	1/12 = 0.08	$log_2(0.08)$ = -3.64	0.08*-3.64 = -0.29
BLACK	1	1/12 = 0.08	$log_2(0.08)$ = -3.64	0.08*-3.64 = -0.29
TOTAL				-1.4

	C	PI	LOG_2PI	$(PI)(LOG_2PI)$
SPECIES RICHNESS (S)	4			
WHITE	3	3/12 = 0.25	$log_2(0.25)$ = -2	0.25 * -2 = -0.5
DARK GREY	3	3/12 = 0.25	$log_2(0.25)$ = -2	0.25 * -2 = -0.5
LIGHT GREY	3	3/12 = 0.25	$log_2(0.25)$ = -2	0.25 * -2 = -0.5
BLACK	3	3/12 = 0.25	$log_2(0.25)$ = -2	0.25 * -2 = -0.5
TOTAL				-2

The Shannon-Weiner index then reverses the polarity, so the results we get are:

Population a) H′ = 0

Population b) H′ = 1.4

Population c) H′ = 2

This calculation uses a higher reading to show a higher level of diversity in the studied population.

Evenness

One more calculation using the same numbers is for evenness or E. I think we can guess which population is the most even out of the three ball pits, but we can prove it with maths.

The equation this time is:

$$E = {}^{H'}/_{Hmax}$$

Hmax is $\log_2 S$ where S is the number of species in the community, which we calculated earlier. $\log_2$ means log to the base 2.

For population a) E = 0/ $\log_2(1)$ = 0/0 = 0 i.e. no evenness at all!

For population b) E = 1.4/ $\log_2(4)$ = 1.4/2 = 0.7

For population c) E = 2/ $\log_2(4)$ = 2/2 = 1 or 100% even.

From our simple population, we can see that the monoculture population a) has no diversity or evenness at all, whereas population c) is the most diverse and the most even. We can think of day-to-day examples where we know this to be true. In a garden or allotment where vegetables are grown, gardeners often plant a diverse range of plants: they know that a good year for one crop is likely to mean a bad year for another, as different plants thrive in slightly different conditions.

Hill numbers

From the above we can work out the evenness of a population. However, there are several hundred different calculations and equations that could be used. Another example is known as the Hill number. This better reflects the **effective number of species** needed and are often used when looking at the conservation of organisms within ecosystems.

The equation is:

$${}^{q}D = \left(\sum_{i=1}^{S} pi^{q}\right)^{1/[1-q]}$$

For this equation, q= 0, 1 or 2.

When q = 0, the equation becomes:

$${}^{0}D = \left(\sum_{i=1}^{S} pi^{0}\right) = \mathrm{S}$$

From previous calculations, we know that S is the number of species.

When q=1, then equation becomes e^H or:

$${}^{1}D = e^{H} = e^{(-\Sigma pi \log pi)}$$

Where H is the Shannon-Weiner index.

When q=2, the equation is in effect 1/Simpson index we saw earlier.

As q increases, rare species have decreasing weight in the overall result of the equation. The common species have a relatively strong impact on the index.

Summary

Biodiversity is vital to the ongoing survival of the human race. There are more than 200 different calculations of diversity available, of which we have seen several in action. It is important to have an accurate picture of the species type and variety prior to human interaction, whether it be planting crops, planning a housing scheme or removing a forest. Once a species is lost from an area, it is very difficult to replace and the entire food web can be affected.

Chapter 11 - The soil ecosystem

A nation that destroys its soils destroys itself.
- Franklin Delano Roosevelt

Soil sounds pretty boring. What could possibly be important about the brown stuff underneath our feet? However, soil is probably the most important of all the ecosystems. Without good soil health the planet will rapidly end up in trouble. The phrase 'the life in earth drives the life on earth' is apt, as we will see.

The soil is known by some as the 'factory of life', although many would struggle to mention any of the workers who live there or the jobs that they do.

The soil provides the abiotic and biotic factors needed for land-based life to thrive. The lithosphere is the portion of land that contains rocks, sand and soils.

Sand, silt and clay

To start, we need to understand the basic constituents of soil. Soil can be described as sand, clay or loam or a combination of the three. It can be represented in a triangle, as Fig. 1 shows. To read the chart, follow the percentage line that is horizontal to the baseline for that constituent.

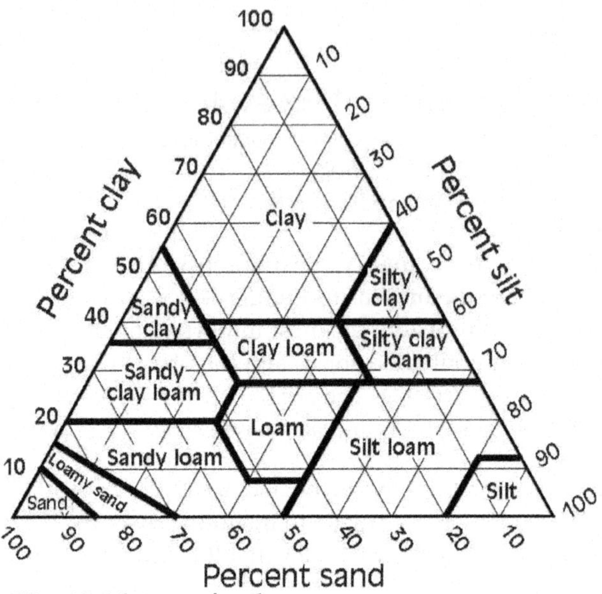

Fig. 1 Make-up of soil

The difference between silt, sand and clay is down to the size of the particles. This is demonstrated in Fig. 2.

Fig. 2 Particle size of different soils

To identify the make-up of a soil sample, fill about one-third of a jam jar and top up with water. If shaken and allowed to settle, the three layers become easier to see. Heavier sand particles sink to the bottom quickly - often within a minute or two. Finer silt particles can take a couple of hours to settle into a discrete layer. The finest particles form a clay layer and takes until the water has cleared.

We can combine Fig. 1 and Fig. 2 with our own separation test. For example, if a soil has 30% clay-sized particles, 10% silt-sized particles and 60% sand-sized particles, then it would be called a sandy clay loam. If there were 30% sand and 70% silt, it would be a silt loam.

Depending on the percentage of clay, silt and sand, each soil type has different characteristics. Soils with a high clay content have good water- and nutrient-holding capabilities, but drain poorly. Equally, sandy soils drain well, but water and nutrient retention are poorer.

This can be imagined with golf balls representing sand particles, marbles representing silt particles and grains of salt representing clay particles. Find a two-litre plastic bottle, remove the top and replace the bottom of the bottle with a fine mesh. If the golf balls are added and water poured through, the water will drain through quickly. If the golf balls and marbles are added together, they will fit closer together and there will be less space for the water to pass. These are called 'pore spaces'. If the golf balls, then the marbles, and finally the salt are added, there will be far fewer pore spaces or air in our modified bottle. As a result, water will pass or drain more slowly and more water will be retained.

Soils with large pore spaces such as sands drain water far more quickly; water is influenced more by gravitational pull than capillary action. Capillary action influences small pore space soils such as clays. Gravitational pull works faster than capillary action so water drains more easily from sands than clays. Organic matter (see below) is better at holding water and nutrients than sandy soils, which is important for plant growth. Equally, removing plants from an area leads to less organic matter, creating sandier conditions and exacerbating the cycle of the land's fertility decline.

Soil depth

Just as children who play Minecraft know, if you dig deep enough through soil you eventually reach bedrock. Unlike in Minecraft, in the real world, in different locations you need to dig to different depths to reach rock.

Soil can be split into layers. These are labelled as organic material (O), topsoil (A), subsoil (B), weathered rock (C) and bedrock (R) in Fig. 3. There can be an additional layer - E - between A and B, but this layer tends to occur only in forested areas. Not all soils have all layers. The topsoil often has a far higher organic material content and, over time, fewer clay particles. The subsoil tends to accumulate more minerals, salts and clay particles. More organic matter accumulates in soils from grasslands than from forests.

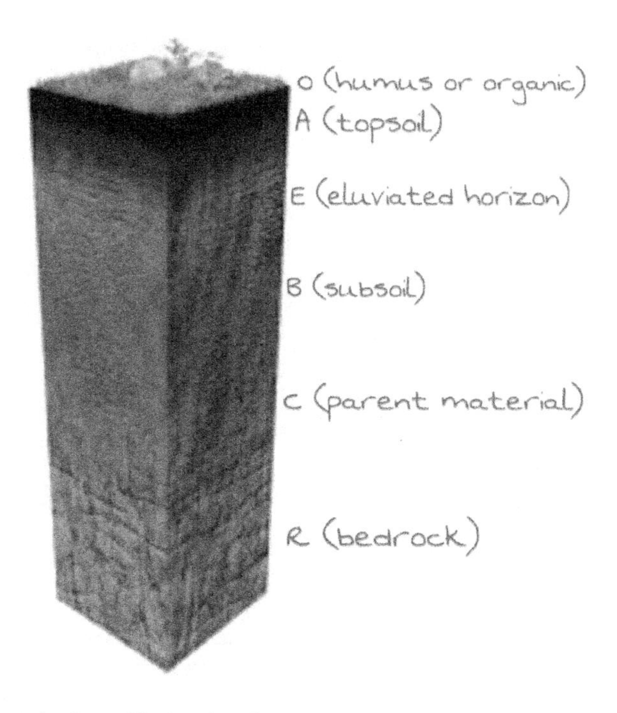

Fig. 3 Soil profile by depth

The structure of the O, A, B and E layers is influenced by the soil texture and composition, the activity of organisms such as worms within the soil, moisture levels, soil compaction and freezing/thawing cycles.

Temperature, nutrient levels, amount of organic material and water levels all vary by depth and also by time of year and soil type. Different plant root systems can extend to different depths.

Temperature

Towards the surface, the land is heated by the sun. This leads to a higher temperature and is obviously seasonal. Wet soils tend to retain heat slightly better than dry soils.

By about 15m down, the temperature levels off to a steady 10°C in the UK regardless of season. This knowledge is used for some underground water-heating systems.

Biogeography

The importance of soils in different regions was beautifully demonstrated by Fierer and Jackson in 2006. Their research sampled soil at 98 sites throughout North and South America. They divided the soil types into one of six categories and analysed the bacteria present using their ribosomal DNA. The six types were: tropical forest/grassland, snow forest (known as a boreal forest), temperate forest, dry forest, temperate grassland and dry grassland. Bacteria were found everywhere, but different species were seen in different places and the number of species varied. Overall, they were able to show, using an extension to the diversity calculations we saw in Chapter 10, that it was soil acidity that had most impact. In a soil that was more acidic there were fewer types of bacteria present. Surprisingly, temperature *didn't* change the number of species of bacteria in the soil and nor did amount of water (measured by potential evapotranspiration), although these factors *did* have an effect on the plant and animal species in each area.

An abiotic factor: the carbon cycle

Nutrient recycling is primarily done by microorganisms. We have seen the importance of carbon, nitrogen, phosphorus and sulphur along with oxygen in the building blocks of DNA, amino acids, proteins, fats and carbohydrates. Our cells contain a complex mixture of these and other elements.

We saw the nitrogen cycle in Chapter 10 and now turn our attention to the carbon cycle.

Carbon is vital to life: it is in every cell of every living organism on the planet. The simplest version of the carbon cycle shows plants turning carbon from the air (in the form of carbon dioxide; CO_2) into sugars that form plant cells. This is photosynthesis and is done by the chloroplasts in the plant cells. We also saw mitochondria making carbon dioxide during the creation of ATP in Chapter 4 during respiration. The carbon dioxide is passed to the atmosphere and forms a cycle, as in Fig. 4.

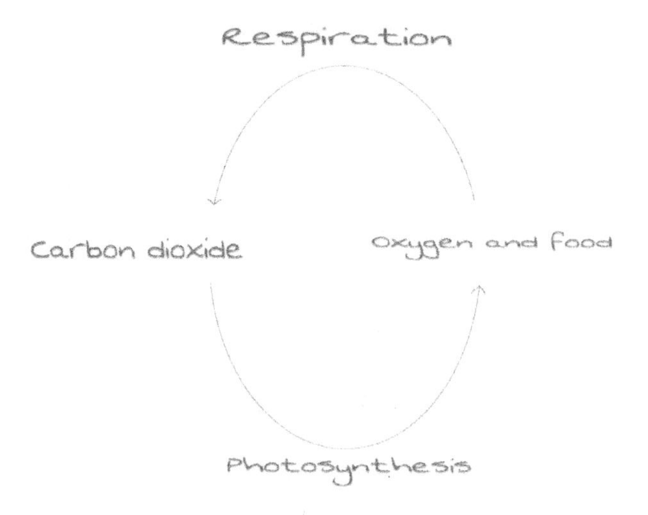

Fig. 4 Simple carbon cycle

Chemically, this can be written as:

Respiration:

$C_6H_{12}O_6$ (organic matter) + $6O_2 \rightarrow 6CO_2 + 6\ H_2O$ + energy

Photosynthesis:

energy (sunlight) + $6CO_2 + H_2O \rightarrow C_6H_{12}O_6 + 6O_2$

This cycle can continue eternally in a closed system, such as the terrarium in Fig. 5, as long as there is a balance between the oxygen produced and the oxygen consumed. When all that is produced is being consumed, the cycle repeats itself.

Fig. 5 Artificial terrarium

Plants create sugars as carbohydrates and ultimately more plant life by photosynthesis. Importantly, this only happens during the growing season and is responsible for the seasonal variation of atmospheric carbon dioxide levels that are seen on the Keeling curve. This is a graph of atmospheric carbon dioxide measured on a Hawaiian island and varies from summer to winter. The plants are eaten by the primary consumers seen in Chapter 10, and both plants and animals respire carbon dioxide.

Over the history of the Earth, many plants and animals have come and gone. As some of the plants and animals died, the carbon became 'trapped' as their bodies were buried and decomposed. Over millions of years and with immense pressures, carbon-rich fossil fuels such as coal and oil were created. Humankind found that fossil fuels were an excellent store of energy and started to burn them. Since the Industrial Revolution, more fossil fuels have been dug out of the ground and burned. During combustion, the carbon combines with oxygen from the air and a lot of carbon dioxide is released. As a result of the extra carbon dioxide, the balance between the carbon that can be used by plants and the carbon released by the combination of normal respiration, in addition to that released by humankind, has been affected. Overall, there is about 30% more carbon dioxide in the atmosphere than there was 150 years ago - the highest levels for nearly half a billion years.

Microbes in the soil decompose both plants and animals, adding to the organic carbon content of the soil; they also respire and generate carbon dioxide. Some species combine carbon with hydrogen to generate methane (chemically CH_4).

The carbon cycle involves many different players. In Fig. 6 we see a more complex carbon cycle.

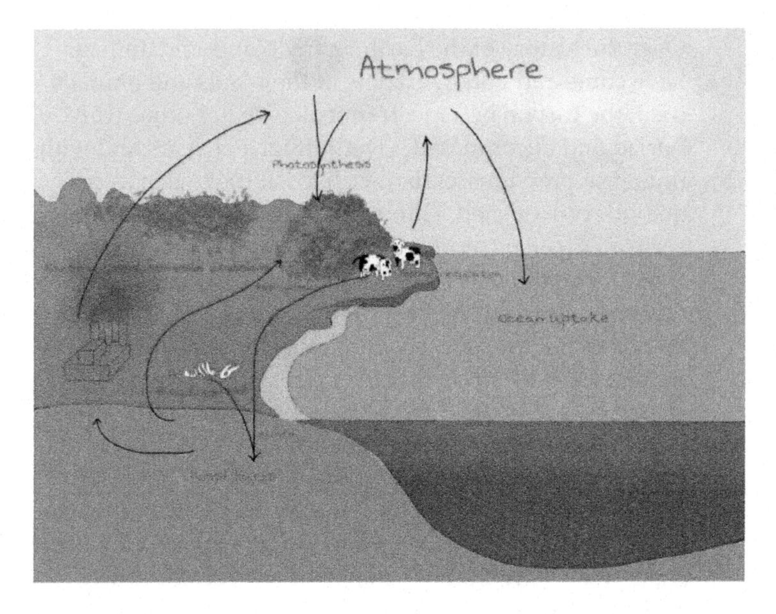

Fig. 6 Carbon cycle

The carbon stored within the earth, whether on land or sea, is known as a 'sink'. Dead plants and animals being buried on the land or sinking to the depths of the oceans 'locks' away the carbon. Carbon dioxide is also dissolved in the water of the oceans. A human is a sink of sorts - over half our body (excluding water) is carbon!

If more carbon is released than absorbed, then it is called a 'source'. The carbon cycle is a balance between sinks and sources, as we can see in Fig. 7.

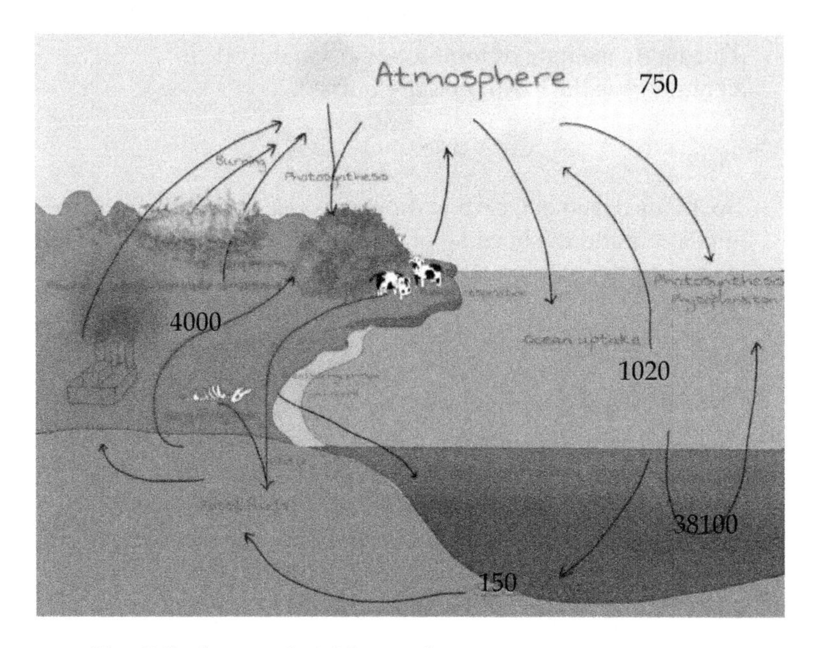

Fig. 7 Carbon cycle with numbers

The black numbers in Fig. 7 refer to the number of gigatons (that is 1,000,000,000 tons!) of carbon dioxide. We can see that carbon moves from one place to another - from source to sink and back again.

Out of balance?

Humankind is producing carbon dioxide at a greater rate than it can be removed from the atmosphere. Burning fossil fuels and creating concrete both release carbon dioxide. Making concrete actually gives a double whammy. It starts life as limestone or calcium carbonate (chemically known as $CaCO_3$). When heated, calcium carbonate releases carbon dioxide to create lime (calcium oxide or CaO). When calcium oxide is mixed with other chemical compounds and then water, it forms cement.

During the heating of limestone carbon dioxide is generated as the fossil fuels are burned.

$$CaCO3 \rightarrow CaO + CO_2$$

So, to make cement, carbon dioxide is released from the limestone and the burning of the fuels to generate the heat.

Biotic factors: the workforce

Soil has many different organisms involved in the factory. They range from single-celled bacteria up to complex mammals such as moles. They can be divided by size: microfauna, such as bacteria, fungi and protozoa; mesofauna, such as tardigrades and mites; and macrofauna, such as insects, ants, moles and birds.

Bacteria

Bacteria are vital to soil and are probably the most essential component. Some bacteria, such as *Streptomyces coelicolor*, we may recognise if not by name then at least by odour. Their smell is described as the 'smell of warm soil after rain'; the chemical responsible is called geosim. Humans can recognise very low concentrations - even as low as five parts per trillion! However, there are many more species of bacteria in soil; it is estimated that there are 4,000 different species *per gram* of soil, with between 100 million and one billion bacteria per teaspoon.

Growing bacteria found in soil is notoriously difficult. About 90% have yet to be grown in the laboratory. As a result, bacterial RNA is used to identify which species are present, although this method is problematic. We can identify what different species are there, but it is impossible to know what proteins they have or how they work and function. One possible reason why it is difficult to grow bacteria in a laboratory is that they often live in nutrient-poor surroundings and spend a lot of time in a resting state.

Increasingly, some bacteria have developed a working relationship with other organisms, whether other bacteria, fungi, plants or animals. These relationships are called mutualistic, where both parties benefit. Some examples include the bacteria involved in the nitrogen cycle we saw in Chapter 10, such as *Rhizobium* and *Frankia. Rhizobium* form nodules on the roots of legumes such as peas and beans, whereas *Frankia* are found on trees including hazel and alder. Both bacteria use nitrogenase enzymes to fix atmospheric nitrogen into ammonia for the plants to use.

Some bacteria are pathogenic, causing diseases and harm to plants and animals. *Bacillus anthracis* is the bacterial agent causing anthrax. Anthrax has been used as a biological weapon, but causes symptoms in most animals as well as humans. The spores can survive in soil for centuries only to be reactivated when the circumstances are right. Without treatment, 85% of humans who are infected with anthrax die, often due to the powerful toxin produced by the bacteria. Fortunately, there is both a preventative vaccine and an effective treatment.

Tetanus is caused by the soil bacteria *Clostridium tetani*. This disease 'only' kills 10% of people infected but causes muscle spasms and 'lock jaw' where the muscles of the jaw 'lock'. The symptoms are due to tetanospasmin, a highly potent toxin produced by the bacteria. In Chapter 8 we saw that the tetanus toxoid vaccine used inactivated toxin.

Other soil bacteria that can cause potentially fatal human diseases include *Campylobacter* or *Bacillus cereus* (gastroenteritis); *Clostridium botulinum* (botulism); other species of *Clostridium* (gas gangrene); and *Listeria monocytogenes* (infections).

Some bacteria are beneficial and able to keep other bacteria in check. This can be done by providing competition for the available food and nutrients, by becoming parasites and feeding off the bacteria or by producing chemicals that interfere with their growth – such as antibiotic-like compounds. *Pseudomonas fluorescens* can produce a chemical that promotes plant growth and produces a chemical that stops certain plant-damaging fungi from growing. The bacteria, along with other *Pseudomonas* and *Xanthomonas* species, produces chemicals that slow or stop the growth of damaging bacteria or by reducing a pathogen's ability to invade the plant. The antifungal activity of *Pseudomonas* can be very useful to farmers of certain crops in these circumstances.

Other bacteria are decomposers that consume carbon and convert the energy in soil into forms of energy useful to other parts of the food chain.

Cyanobacteria are clever bacteria that can use carbon dioxide to gain energy from the sun during photosynthesis. *Cyanobacteria* are found in most environments on the planet, usually in the top 1-2mm of the soil. As well as photosynthesising, *Cyanobacteria* also fix nitrogen, making them doubly useful; rice fields, for example, are more productive when 'infected'. Bacteria also improve the structure and fertility of the soil and add a level of water-carrying capacity.

Yet another amazing bacterium is *Pseudomonas syringae*. Some plants produce chemicals that are natural antifreezes and stop their cells being damaged during a frost. Some plants keep their cells intact at temperatures as low as -12°C by keeping their tissues as a 'supercooled' liquid. However, if a plant is infected with this bacteria, it freezes at a relatively high -1.8°C, so a relatively mild frost would cause much more damage to the plants. Ice cells form by ice nucleation around a small particle such as a dust granule. A single gene on the DNA of *P. syringae* produces proteins that arrange themselves in such a way as to mimic the crystals found in ice, which means that more ice forms around them.

Algae

Algae are bacteria that can photosynthesise and are found near the surface of many soils. Surprisingly, some algae have been found nearly 20cm below the surface; it is thought they travel there by riding on earthworms or by being carried by rainwater trickling down. Soil algae include the green algae *Chlorophyta*, yellow-green algae *Xanthophyta* and red algae *Rhodophyta*.

Lichen

Lichens are a combination of algae and fungi that live and work together. They produce their own food by photosynthesis and are incredibly slow-growing. Some are discussed in Chapter 13.

Fungi

Fungi are known to many as mushrooms, but there are far more varieties living, surviving and thriving within the soil. They can often be found well below ground and their 'roots' or hyphae can form huge networks called mycelium. In just one square meter of grassland, there can be more than a kilometre of fungal hyphae! *Armillaria bulbosa* is a fungus found in forests that can have a network of mycelium weighing several hundred tons and stretching several kilometres. The largest mycelium was found in Oregon, USA, and is thought to be 890 hectares or 2,200 acres in size. The dense network of mycelium stabilises the soil and holds it together.

Fungi are involved in the growth of plants. Over millions of years they have formed cooperative, symbiotic relationships with each other. As a result, many trees cannot grow without fungi helping their roots to get hold of the nutrients they need. A common group of fungi is arbuscular mycorrhizal fungi (AMF). This magical little fungus is found in over 80% of plant species. It invades the roots of plants and then sends its own long hyphae into the surrounding soil, as seen in Fig. 8. In soil, this can look like tiny white threads as if from a spider's web. There are more than 200 fungal species that help plants by absorbing phosphorous and nitrogen (as ammonia) from the soil while being given carbon by the plant. The loss of plants and fungi has a big impact on the structure and stability of the soil.

In addition to helping plants gain nutrition from the soil, fungi can also break down almost all organic materials. As a result, they are known as primary decomposers. Their enzymes can break down complicated structures such as wood.

The most expensive food on the planet is a fungus: the Perigord truffle (*Tuber melanosporum*). In 2014, a truffle smaller than a football and weighing 1.7kg was sold for a staggering $61,250 or £42,500!

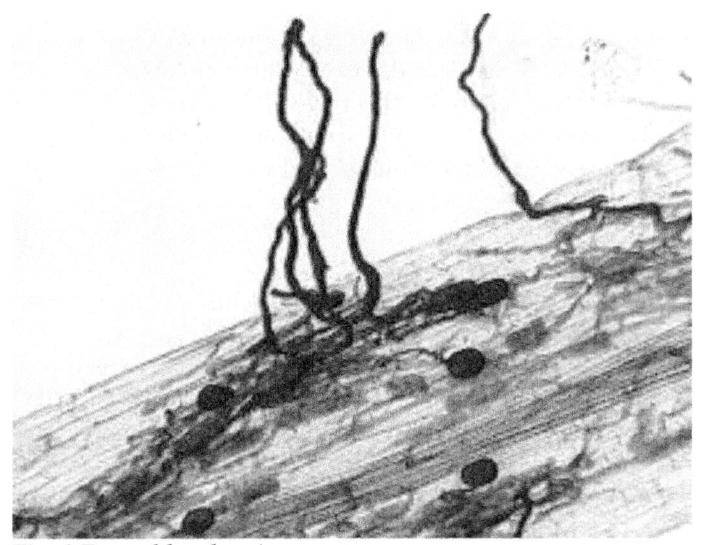
Fig. 8 Fungal hyphae in a root

Unfortunately, not all fungi are good for humans. There are many examples of deadly fungi such as the death cap (*Amanita arocheae* or *Amanita phalloides*), destroying angel (*Amanita bisporigera* or *Amanita exitialis*) and deadly webcap (*Cortinarius rubellus*). These tend to produce deadly chemicals such as amatoxins, ergotamine or orellanine, which, when eaten, cause symptoms such as heart, liver or kidney failure. The growth of a fungus such as ergot (*Claviceps purpurea*) on grain or rye causes hallucinations a bit like LSD. It has been implicated in the purported mass hysteria of the Salem witch trials or when mass poisonings or hallucinations affect whole villages or communities.

Slime moulds

Slime moulds look like fungi and some behave like fungi but are actually entirely different. There are two main groups: myxomycetes and dictostelids. Both eat dead material - especially decaying organic matter such as leaves and trees - but also eat bacteria and algae. In part due to their small size, this single-celled group of organisms has not been well studied or understood. Fascinatingly, slime moulds have some 'intelligence' and can find the shortest way around a maze!

Protozoa

Protozoa, like slime moulds, are also made of a single cell and eat bacteria. There are more protozoa - over a million per teaspoon - in fertile soils. Different soil types have different protozoa species present; more than 30,000 species have been discovered so far. The best-known protozoan is probably the amoeba, although they can come in an impressive range of shapes and sizes. As they feed, nitrogen and other essential nutrients are released into the soil. The nitrogen is in the form of ammonia and is used by higher-level consumers in the food web.

Small animals

Small animals such as tardigrades ('water bears'), rotifers (miniature leach-like creatures) and nematodes (small worms) all live in the soil. They feed on bacteria, fungal cells, protozoa and on some plants. There are vast numbers of each of these species, with more than 1,000 types of tardigrades, about 460 types of rotifers and more than 30,000 types of nematodes discovered so far. It is likely that there are many more waiting to be discovered.

Some nematodes are harmful to animals and humans, such as hookworms and pinworms. In dogs, heartworms (*Dirofilaria immitus*) cause a disease affecting the heart and blood vessels. However, humans have put other nematodes to good use above and beyond their nutrient cycling. One nematode (*Steinernema*) helps to control insects that would otherwise damage crops. It does this by entering the insect and infecting it by releasing bacteria from a special storage pouch. The bacteria then release a toxin to kill the insect, so providing food for the nematodes' young.

Insects

Insects of all sizes - ranging from 0.2mm soil mites (*Acari*) to 0.5mm springtails (*Collembola*) or potworms (*Annelids*) up to 20mm long - live in the soil. These can help with pollination of plants and mosses, cycling of nutrients, and eating dead and decaying matter. Some feed on smaller organisms, others on nutrients in the soil itself.

Larger insects such as earthworms or multi-legged creatures such as millipedes and other arthropods make the soil home. Life can be difficult - for example, for an earthworm to get enough calories, it needs to chew through 10-30 times its own body weight of soil each day; as some earthworms are over one meter long, that is a lot of soil! Unfortunately for the worm, they don't have the necessary enzymes to digest plant material so rely on bacteria to do the digesting for it. Earthworms are vital for transport of water and nutrients through different soil layers as they can form a burrowing network up to three metres in depth. These burrows are used by plant roots to explore the depths of the soil layers. The number of earthworms in a field of maize could be 20 per square meter, while in a meadow is closer to 300 per square meter.

Ants

Ants are related to bees and wasps and are from the same order: *Hymenoptera*. More than 12,500 species of ants have been found on plants so far. Amazingly, all the ants together account for between 15-25% of the mass of all animals combined. Ants provide many functions for the soil including being predators, scavengers or mutualists. They can help disperse seeds and can 'farm' some plants, animals or fungi for food. For example, ants are the protectors of aphids. By protecting an aphid from a predator, the ants are rewarded with a sugary substance called honeydew from the aphid, which the ant uses for food. Ants have two stomachs, one for their own food, and one to store food for other ants. Ants are a keystone species in some soil ecosystems and can be a good marker of pollution levels.

Many, many more varieties of animals exist in the soil. Beyond the above, other animals include beetles, woodlice, termites, spiders, fly larvae, slugs and snails, frogs, moles and birds.

Regulation of plant communities

Plants have been on Earth for 400–450 million years. They evolved from bacteria that could use energy from the sun to power their chemical reactions and specifically generate ATP. The endosymbiosis theory points to a photosynthesising bacteria being consumed by a prokaryotic cell and eventually becoming a chloroplast. Plant life would initially have originated as water-dwelling life. Co-evolution with certain fungi subsequently allowed early photosynthesising eukaryotic cells to conquer the land.

In 2016, a report from Kew Gardens estimated that there were about 400,000 plant species known to humankind, with about 2,000 new species discovered annually. Unfortunately, more than 20% of these plants are at risk due to climate change, other invasive species, habitat loss and diseases.

Uses of soil ecosystem

The broad range of functions of soil can be summarised as 'ecosystem services'. Soil is the sum total of the abiotic and biotic factors and drives life on Earth. It supports soil creation and organic matter, provides plants and crops and cycle nutrients. It provides biodiversity and biomaterials and foods. It is able to regulate erosions, water quality and gas quality and regulate the climate. It also provides cultural benefits including recreation and heritage.

Threats to soils and plant life

We have seen how the soil type influences what plants and animals exist; in turn, the plants and animals influence the soil type that exists. Threats occur when one of the parts of the cycle is disrupted and has a knock-on effect on the other parts of the cycle. For example, clearance of woodlands or forests reduces the amount of organic material, such as leaves, that is available. The organisms that rely upon the organic material for their food to produce energy are then unable to survive. This affects the cycling of nutrients into the soil, leaving it nutrient-poor. The result of that is being less able to grow other plants without adding artificial fertiliser, which has a further effect on the organisms that thrive.

Other influences include changing weather or climate patterns - for example, more or less rainfall. Soil moisture levels are affected, subsequently changing the plant or animal species able to survive. Diversity of species is also affected by erosion of the soil, falling soil stability by removing root systems, or changes to the management of the land such as intensive grazing or planting of monocultures.

Human impacts on soils can include mechanical compaction, which reduces pore size between the soil particles and the subsequent effect on moisture retention. Chemical fertilisers affect the balance of microbes by encouraging some species to reproduce faster than others and upsetting the natural balance. Concreting or tarmacking land obviously seals the soil and allows roads and housing to be built, but causes water to 'run off' rather than soaking into the soil. Pesticides remove some key species from complex food webs. If these species are 'keystone species', the effect is even more profound. For example, if decomposers or nutrient cycles are predominantly affected, the nutrient composition of the remaining soil is affected. Equally, if pollinators such as bees are removed, then the plants or crops cannot be pollinated. Some crops need very specific bees to pollinate them - for example, the squash bee (*Peponapis pruinosa*) only collects pollen from squashes and pumpkins and often nests on the ground nearby. Cacao (the cocoa bean) is pollinated by small flies in 90% of cases; the flies are dependent on the cacao as they reproduce on the surface layer of the soil made from decaying cacao pods.

Salinisation of soil occurs when the freshwater supplies are used - especially for irrigation and agriculture or extraction for human consumption - leaving less water and more salts behind.

Invasive plants are non-native plants that move across the world and can cause many problems by out-competing native plants for resources and nutrients. Examples in the UK include Japanese knotweed or giant hogweed. The UK government takes the problem seriously; if someone allows the weed to spread in the wild, it can carry a fine of up to £5,000. Pathogens can also be transmitted around the world such as the fungus threatening ash trees we saw in Chapter 10. Microorganisms can damage crops such as the fungus (*Gaeumannomyces graminis* var. *tritici*), which attacks the root system of wheat plants. Plants and seeds can be transported across continents on the feet of migrating birds. Some invasive plants are for sale on websites such as eBay. Some states in America - notably California - are very strict on the movement of plants, so fearful are they of diseases devastating their local economy.

Future needs for biodiversity

The incredible range of microorganisms in the soil provides a vast array of chemical pathways, most of which are unknown or unexploited so far. It is almost like alchemy to turn one substance into another, but this is precisely what biotechnology hopes to achieve. First, the bacteria need to be identified and cultured during bio-mining. Culturing is usually done initially in a petri dish in the laboratory and then in giant bioreactors where the conditions can be controlled. Bioremediation uses decomposers to clean up soils by breaking down undesirable chemicals into less harmful ones. This also allows the recycling of nutrients and has been done very successfully in some areas contaminated with oil by-products. Bacteria can clean up metals such as zinc, mercury and cadmium from waters and waste from mines. Other bacteria clean up organic compounds and convert them into carbon and energy.

Fungi are used in the production of Quorn as mycoprotein, but the mycelium has also been used to make biodegradable packaging, and a company called Ecovative uses it to make surfboards.

Bacteria such as *Agrobacterium radiobacter* and *Achromobacter* sp. are used in the treatment of wastewater, while wood-rotting fungi such as *Phanerochaete chrysosporium* can break down solid waste from agriculture.

Bacteria can be used for pest control; this is termed biocontrol. They can be targeted to specific insects or pathogens rather than a broad approach that affects all the wildlife in the sprayed or treated area. Small organisms such as nematodes have been used as pest controllers.

Antibiotics are discussed in Chapter 8. We saw that soil bacteria are a rich source of antibiotics and that it was from the fungus *Penicillium glaucum* that penicillin was obtained.

Biogas and biofuel production are discussed in more depth in Chapter 14.

Summary

Healthy soils are vital to a healthy planet, but soil nutrition is being depleted significantly faster than it can be replenished, mainly through unsustainable farming methods. According to estimates by the United Nations Food and Agriculture Organisation, about one-third of total food production is wasted each year. Globally this equates to about 1.3 billion tonnes of food; continuing to plant food crops and then not use them is a travesty. Globally, the increased demand for meat is accelerating deforestation as the land is cleared to grow plants as feed for livestock. World hunger will persist and soil quality will deteriorate.

Soil bacteria have been used to create antibiotics and other important chemicals. The planet damages the soil ecosystem at its own peril.

Chapter 12 - The marine ecosystem

A society is defined not only by what it creates, but by what it refuses to destroy.

– John Sawhill, former president of The Nature Conservancy

Aquatic ecosystems

The Earth is covered with water. About 71%, or 360,000,000m^2, of our blue planet is water. In scientific notation, this is written as $3.6 \times 10^8 m^2$. In total, there are 1.4×10^{21} litres of water; 97% of this is in the oceans. This makes it by far the largest single biosphere. We have seen the huge impact it had on how life began and will discuss how it will continue to impact on what happens to humankind in the future.

There are many different parts to the aquatic ecosystem; they can be split into freshwater and salt water or marine.

Freshwater makes up only about 2.75% of the total amount of water on Earth. Of that, between 1.75% and 2% is frozen either in glaciers, snow and ice. Another 0.5-0.75% is groundwater and soil moisture. A further 0.01% is in rivers, swamps and streams, but the majority is in freshwater lakes. Overall, less than 0.003% of water on Earth is fresh and unpolluted.

Marine environments

We will look mainly at the salty environments as salt water covers the majority of the globe. Marine settings can be broken down into different regions depending on how close they are to land and whether there is fresh water and salt water present, but first we will look at the tides.

Tides are generated due to gravitational pull from the sun and moon. Although the sun is far larger, the moon has a greater effect on the tides due to its close proximity to the Earth. This 'pull' of the moon causes a 'tidal bulge' as it rotates around the Earth. High tides occur every 24 hours and 50 minutes. There is a constant cycle of high tide followed by low tide at regular 12-hour 25-minute intervals. When the sun and moon line up with each other, there is an additive effect of both gravities together that causes extra-large tides called 'spring tides'. 'Neap tides' are the opposite and are much smaller than average. These occur when the moon is at right angles (or quarter phases) to the Earth so the sun and moon are both pulling the oceans at right angles to each other. The highest tides occur in the Bay of Fundy, which separates Nova Scotia from New Brunswick in Canada. They can measure over 16m – higher than a three-story building.

Where the land meets the sea there are shorelines that can be rocky or sandy beaches making up the littoral zones. This is an area of transition for plants and animals. Intertidal zones are the areas that are affected by the tide. The tides cause deposition of organisms on the beach and removal of waste products. Nutrients are moved and eggs dispersed. Sand is created by the mechanical weathering of larger rocks. As rocks are pounded together, small pieces gradually break off. The minerals within rocks can be dissolved into the water. The type of sand created depends on the origin of the rocks. For example, one type of rock from a volcano is called olivine. This is made from magnesium or iron with silica (chemically either Mg_2SiO_4 or Fe_2SiO_4) and is green in colour. As it is gradually worn down it produces a green-coloured sand, as found on some of Hawaii's volcanic islands. On rocky shorelines there can be tidal pools that offer a protective environment to some organisms.

However, there are environmental conditions when the organisms need to adapt to low water/high salinity conditions due to evaporation or, conversely, low salinity due to rainfall. Rock pools offer a greater exposure to air and warmth due to sunlight.

Estuaries occur and have seawater mixing with salt water. They can be classified as bays, sounds (such as Prince William Sound in Canada, which was affected by the Exxon Valdez oil spill in 1989), mangrove forests, mud flats or swamps. As salt water is denser than fresh water it 'sinks' to the bottom while fresh water forms a surface layer. Therefore, this is a 'layered condition'. There is an increased availability of nutrients in part due to decay of plant life and bacterial breakdown, but there is also an accumulation of mud sediments and other nutrients washed down in the rivers. The tide has an influence in the renewal of salty water. Estuaries offer some protection to the land from storm surges.

Mangrove forests are often made of *Rhizophora* trees. Their roots are able to collect nutrition from sediment but also filter salt from 'brackish' water. These are important to protect coastlines and prevent erosion.

Mud flats tend to receive their nutrients from tidal flow. There is oxygen only to about 1cm depth. As a result, mud-dwelling creatures tend to make breathing tubes through the mud to reach the surface.

Salt marshes have a variable amount of salinity depending on proximity to the ocean, rainfall patterns and river run-off. There is usually a plentiful supply of microorganisms, insects, worms, crustaceans and birds that live in salt marshes.

Oceans
All the main bodies of water are joined to each other so it is hard to determine one boundary from another. However, the body of water traditionally has been divided into five oceans. The largest is the Pacific Ocean, followed by the Atlantic, Indian and Southern oceans. The smallest is the Arctic Ocean, which is frozen apart from around the edges.

There are various characteristics of a cross-section of the ocean. We have discussed the border between the land and sea above. After the shoreline, there is the continental shelf, as in Fig. 1. This extends to where the depth of subsequent water is 100-200m deep, although it can go as far as 320km out to sea.

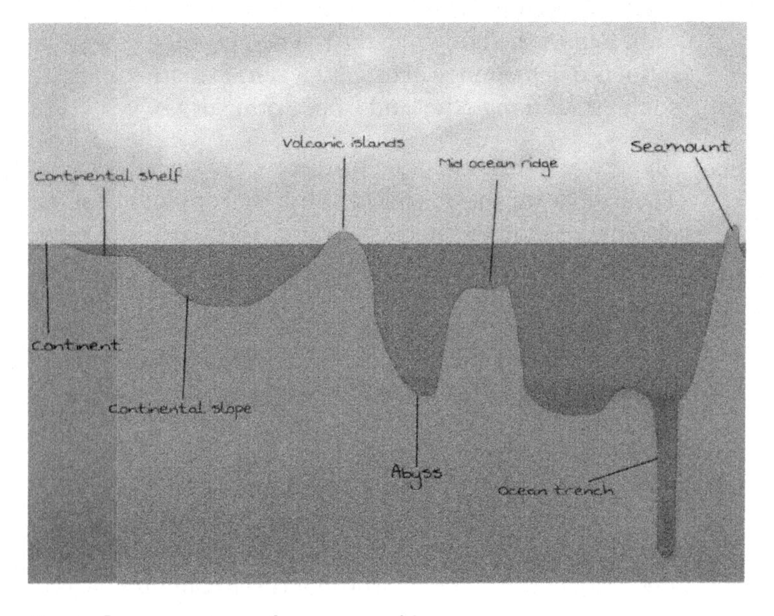

Fig. 1 Cross-section of ocean profile

Beyond this is the deep ocean. The Mariana trench is the deepest part of the ocean, west of Japan and the Philippines in the Pacific Ocean. Here the ocean floor plunges to depths of about 11km. In comparison, Mount Everest is about 10km high! On the seabed, the pressure is about 1,000 times that on land. The trench was formed when the great Pacific tectonic plate was forced under the Mariana plate.

There have been four manned explorations of the Mariana trench, carried out by underwater vehicles, submarines and submersibles. The most recent was in 2012 by film director James Cameron.

Seamounts are another name for sea mountains. These are similar to volcanic islands, although seamounts tend to be smaller. They can rise from the seabed and form small islands. These are a hotspot for various forms of marine life. Commonly, there are coral reefs such as *Lophelia pertusa*. Fish such as orange roughy are fished here and are particularly vulnerable to overexploitation due to their long life expectancy and late sexual maturity.

Hydrothermal vents can be found at the bottom of the oceans. These are fissures in the rocks in volcanically active areas or where the tectonic plates are moving apart. On land, these would be seen at hot springs and geysers. Underwater, they can be seen giving off a black 'smoke' and are unsurprisingly named 'black smokers'. This 'smoke' is a mix of gases such as methane (CH_4), hydrogen (H_2), hydrogen sulphide (H_2S), ammonia (NH_3) and iron (Fe), as in Fig. 2. These areas are teeming with life, although it is estimated that less than 1% of organisms that live around these hot vents have been identified. Rather than using sunlight or sugars to generate energy as we saw previously, chemosynthetic bacteria have unique methods to convert these chemicals into energy to sustain life. From the bacteria, other organisms can enter the food web, including tube worms, clamps, limpets and shrimp.

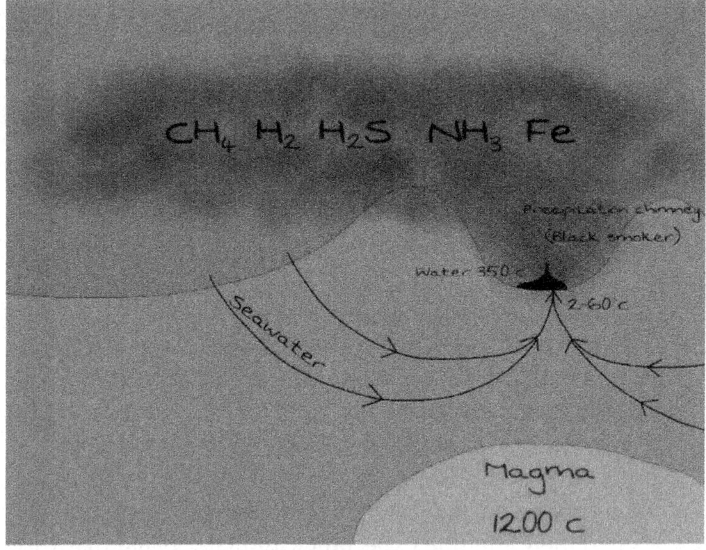

Fig. 2 Chemical factor of a black smoker

It is suspected that life may have started in the hydrothermal vents. It has been shown experimentally that the surfaces of some minerals at these temperatures and pressures act as catalysts to generate organic molecules from the surrounding chemicals. The chemicals of life, such as methane and ammonia, were present and contributed to the formation of the earliest pioneer organism, or Earth Life Form No 1. The extremophiles we met in Chapter 7 would encompass any that we could find at these depths and that could generate energy from sulphur or iron. Günter Wächtershäuser proposed this theory in 1988.

These organisms use unique enzymes to function in such forbidding environments and are thought of as 'extremozymes'. To have a stable enzyme working efficiently in such extreme conditions is a very promising prospect for industrial and chemical processes.

Abiotic factors

We saw in Chapter 11 the importance and variety of abiotic factors in the soil. There are different abiotic factors in the marine ecosphere.

Salt

Salinity refers to the amount of salt dissolved in water. Why is seawater salty? Salt concentration is determined by dissolved minerals such as sodium and chloride in sea salt. Salt exists in water as sodium ions (Na^+) and chloride ions (Cl^-) and is dissolved from minerals in rocks. Freshwater has less than 0.05% dissolved salts (or 500 parts per million). As this increases to up to 3%, the water is called 'brackish' water. Between 3% and 5%, it is saline water, and above 5% it is brine.

Diffusion refers to the 'desire' of ions to be uniformly concentrated and is the movement to balance the ions. Because the ions are charged, they don't pass across a membrane, but water (which is uncharged) can travel across. Water travels to equalise the ions down a concentration gradient in a process called osmosis.

Brine is called a hypertonic solution. This means that if a cell is placed in brine it will lose water from inside the cell to the surrounding solution. This is dehydration. If a cell has a higher salt concentration inside the cell and was placed in a freshwater solution, then it would gain water, causing it to swell up and burst. An isotonic solution is a solution that has the same amount of salt in solution on the inside and outside the cell.

Some organisms have managed the hypertonic, isotonic and hypotonic environments they live in. Salmon is a species that can live in and travel between both freshwater and seawater habitats. We know that cells need to maintain their internal concentration of salts and water and do this by having a membrane resistant to the movement of the charged salt ions. They can then either pump the salt in or out of their cells through special ion channels depending on the environment. Multicellular animals such as fish or birds manage to excrete salt via their kidneys if in a high salt solution. In freshwater, some fish excrete the water and conserve the salt. Salmon are 'euryhaline' ('eury' = wide range, 'haline' = salt) a magnificent word to describe organisms that live in both fresh and salty water. In fresh water, they filter water quickly, lose a lot of urine and take in salt. In salty water they do the opposite -concentrate the urine and excrete salt.

Light

Sunlight is vital to life. The ocean can be sliced top to bottom and labelled by the amount of light it receives, as seen in Fig. 3.

The top layer is called the sunlight zone, euphotic zone ('eu' = well, 'photic' = light) or epipelagic ('epi' - near', 'pelagic' = ocean surface) zone. Sunlight can penetrate to up to 200m, but the depth is dependent upon the amount of silt or algae present. Below this is the twilight or disphotic zone, which can go as far down as 2,000m. Deeper still is the midnight zone. 90% of the ocean is in this slice and there is no sunlight at all.

In the sunlight zone there is plenty of light for photosynthesis. As a result, over 90% of marine life lives in this zone, ranging from single-celled plankton to larger fish species. There are seasonal variations as to the amount of light received in this layer.

Far less light can penetrate into the twilight zone, and it is rare for any plant species to be present. There is some light by bioluminescence, as we will see later.

In the midnight zone there is no light. There are huge pressures and temperatures here. In this zone mainly live animals, although microorganisms that gain energy without sunlight are found here such as bacteria using hydrogen sulphide from hydrothermal vents.

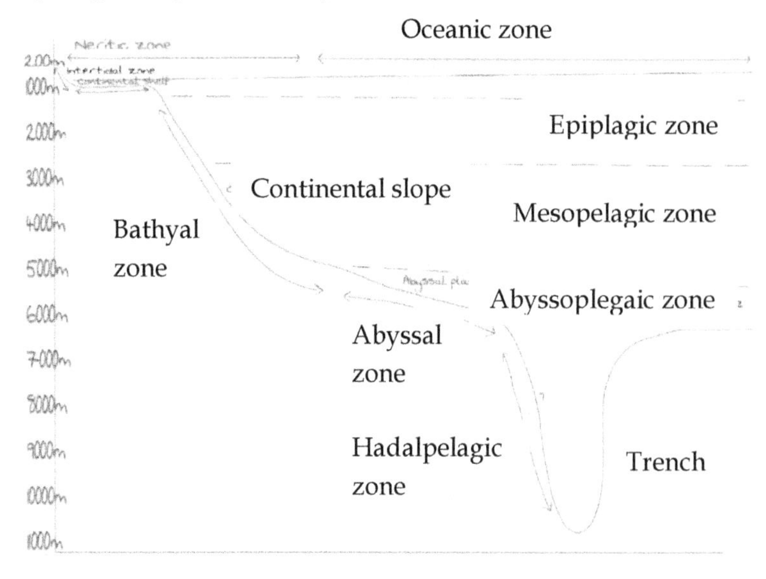

Fig. 3 Marine light zones - horizontal section

Temperature

The temperature falls as the depth of water increases. Unsurprisingly in the sunlight zone, the temperature is higher due to heat transmitted from the sun, and varies by season. The density of water is greatest at 4°C; thus it sinks to the bottom of the ocean as in Fig. 4.

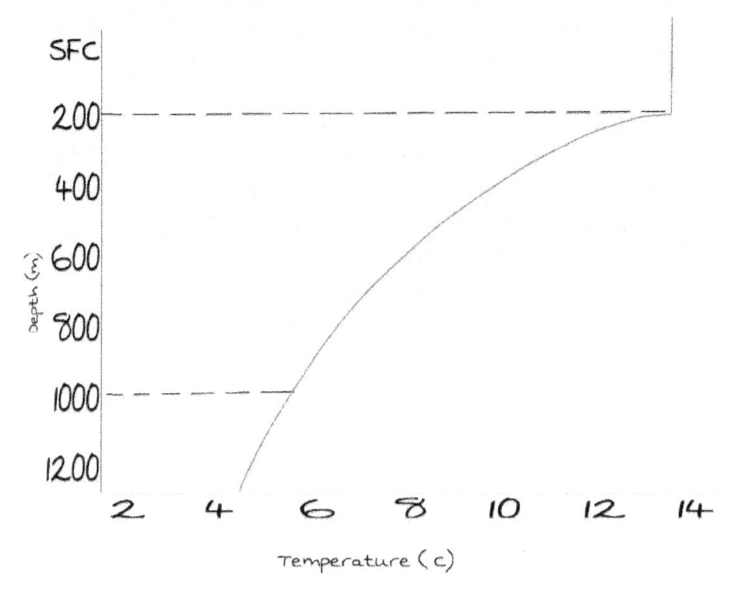

Fig. 4 Temperature changes by depth

Sea temperature also varies by location. As we can see in Fig. 5 the temperature of the sea is colder at the poles than at the equator. However, once you have reached a depth of about 900m the temperature is no longer dependent on the sunshine and is equally cold whether it be summer or winter, pole or equator.

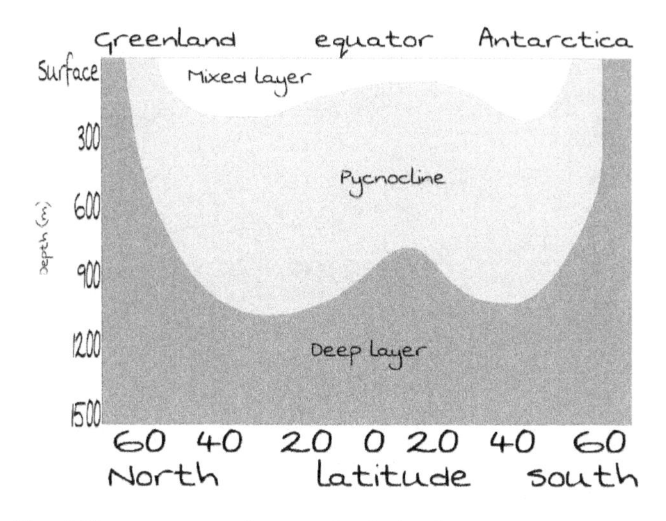

Fig. 5 Temperature changes by latitude

Oxygen levels

Fig. 6 Shows dissolved oxygen levels at different depths. Closer to the surface there is more algae and phytoplankton producing oxygen as well as oxygen dissolved from the atmosphere. As depth increases, oxygen is consumed by organisms during respiration and decay and is referred to as the 'anoxic' or 'without oxygen' zone. Oxygen levels in the ocean are likely to be affected by climate change. 'Dead zones' have no oxygen and are likely to increase in number. Only very few specialised organisms can survive without oxygen, as we saw previously near the 'black smokers'. However, without an alternative energy source, very little if any life will survive at all. Deep oceans can have higher levels of oxygen mainly due to lack of demand and low consumption.

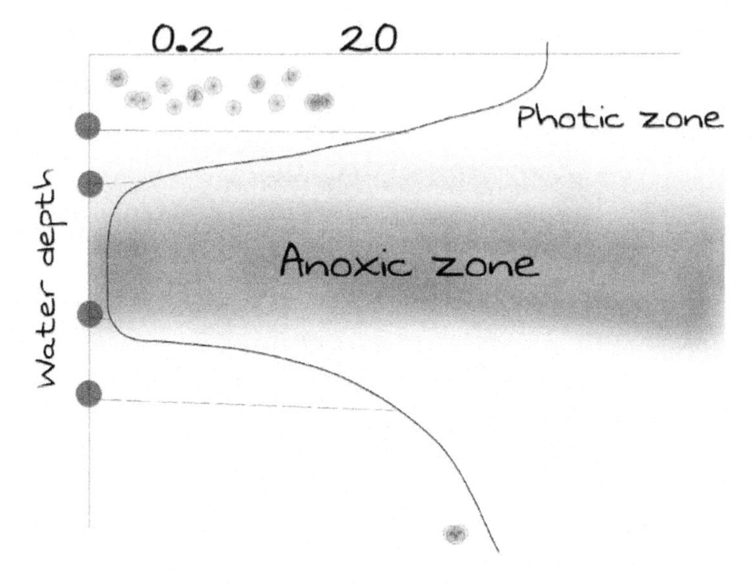

Fig. 6 Changing oxygen levels with depth

Pressure and depth
At sea level, the pressure is known as one atmosphere or one bar. This is the amount of pressure pushing down on us by the weight of the air above our heads. Water is much denser than air and as we go deeper down in the ocean, the more water there is above us – effectively squashing us down. For every 10m we descend, we add an extra atmosphere of pressure. A scuba diver might reach 30m, or three atmospheres. A king penguin (*Aptenodytes patagonicus*) can go as deep as 500m or 50 atmospheres. A submarine may be as far down as 730m, but an elephant seal (*Mirounga leonine*) makes it as far down as 1,500m or 150 atmospheres! Even more impressively, a Curvier's beaked whale (*Ziphius cavirostris*) has made it to nearly 3,000m. Even that pales into insignificance against *Abyssobrotula galatheae* – the

deepest fish ever seen, at 8,370m. The Mariana trench is 10,809m, but life has not been discovered there…yet!

Life was found at the bottom of the trenches, which surprised many due to the immense pressures faced by organisms at such depths. Organisms such as giant-celled amoeba, sea cucumbers and snail fish were seen by using 'dropcams' in 2011 and 2014.

To survive at such depths, different species take different approaches. Elephant seals store extra oxygen in their blood and muscle tissues and have a large number of red blood cells. Curvier's beaked whale has a ribcage that is able to fold down, which decreases air pockets and reduces buoyancy.

Biotic factors

The most important living organisms in the oceans are algae. These single-celled microorganisms are responsible for fixing half of the carbon and removing carbon dioxide from the atmosphere, and for producing half the world's oxygen. Green algae (Genus - *Chlorella*) forms the base of the aquatic food chain. It is easy to grow and can be used as a fertiliser.

The start of the food chain is shown in Fig. 7.

Energy from the sun is converted by algae and other photosynthesising organisms. Plankton refers to a group of organisms including bacteria, algae, protozoa and archaea that cannot swim against the current. Algae and other plankton are fed on by krill - small crustaceans or shrimp-like creatures. Krill (order *Euphausiacea*) are the food source for many other sea creatures, from fish to birds to whales, and are a keystone to the marine food chain. Over the last 30-40 years, stock levels have diminished. Krill gets harvested for aquarium feeds and as bait, and is consumed by humans, especially in Japan and the Philippines. They carry a gene for bioluminescence and so can glow in the sea.

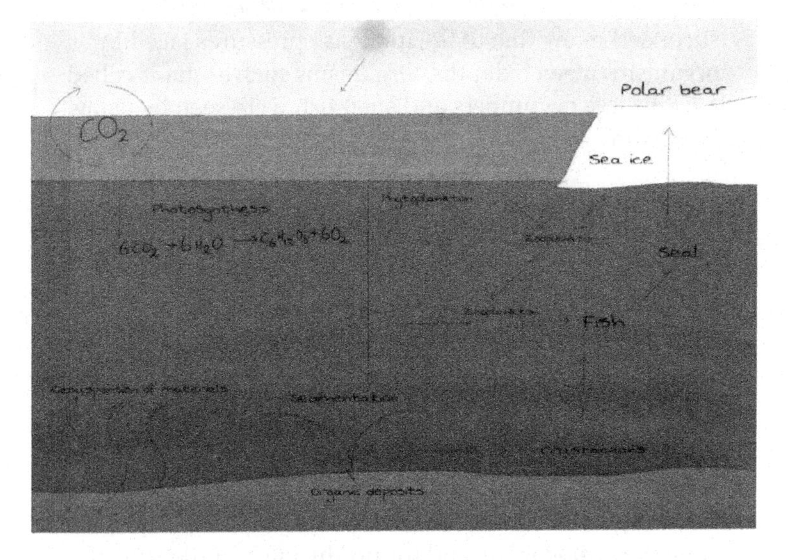

Fig. 7 Distribution of plant life/microorganisms

Other important single-celled algae include *Emiliania huxleyi* - an abundant coccolithophore ('cocco' = seed, 'litho' = stone, 'phore' = bearer) – which, when it forms blooms in nutrient-depleted waters, can kill off krill in large numbers. These organisms are surrounded by calcified scales, which contain a combination of calcium and carbon dioxide and are likely to have an important effect on the outcome of climate change. As the sea becomes more acidic it is possible that fewer of these creatures will absorb the atmospheric carbon dioxide.

Higher up the food web are 'fish'. Fish are the most diverse vertebrates on Earth. As the knowledgeable podcast *No Such Thing as a Fish* demonstrates, there is not really a species of 'fish'. The popular understanding of 'fish' covers at least four groups of vertebrates that are as different from each other as we are from a lizard or a bird. Some fish, such as cod or salmon, have internal bones; others, such as sharks and rays have skeletons made from cartilage. Lampreys and hagfish form smaller groupings and are eel-like with no true backbone. Horseshoe crabs (*Limulus polyphemus*) have been around unchanged for about 150 million years. They are covered in armoured plates that are free to move over one another. Their tail, or telson, acts like a rudder and they can flip themselves right side up if they come to lie on their backs.

The evolution of circulatory systems shows that fish have the simplest heart structure. Theirs consists of two chambers: the atrium holds the blood before it is delivered to the more muscular ventricle to be pumped around the body. Amphibians have two atria and a single ventricle, whereas mammals and birds have a four-chambered heart, with two atria and two ventricles.

Most marine animals breathe through their skin. Tiny blood vessels on the skin surface allow oxygen to enter the bloodstream directly. Examples include sponges, corals and jellyfish.

Birds such as the King penguin feed on small fish and squid. King penguins can be eaten by seals and orcas.

Underwater forests

Just like the forests on land, there are also underwater forests. Instead of being made from trees, they are made of giant algae or seaweed. Algae are single-celled but can still grow up to 40m long! Humans harvest kelp and the protein it produces, alginic acid. This is used in industry as a thickener in many products, ranging from ice cream to polish, hand cream to paint, along with a variety of medicines, gravy, beer, face masks, throat lozenges, imitation dairy products and is even in some sweets; about 50% is used by the textile industry.

Just as in the woodlands we will meet in Chapter 13, the kelp forest has layers. Each profile layer has a top, middle and bottom, and each provides different niches for other sea creatures. Some use kelp as a safe shelter for protection of their young in a safe shelter. Just as with forests on land, kelp forests contain a higher diversity of plants and animals than any other ocean ecosystem. The blades of kelp are held up in the water by bubbles of gas within; this enables them to make the best use of the sunlight for photosynthesising.

Coral reefs

Coral reefs are magnificent structures. Initially the largest was thought to be the Great Barrier Reef in the warm Australian waters. The second largest reef is near Belize in Central America, but a new one, discovered in 2016, has been found off the coast of Brazil and may eventually become the largest reef. Reefs are usually found in shallow, warm waters but there are coral reefs off North-western Europe and in some fjords in Scandinavia.

Lophelia pertusa is a polyp and the main coral builder. Many polyps together form a colony and many colonies form a reef. They are made from calcium carbonate and are very slow-growing.

The corals offer protection for smaller fish from larger predators and form a home to sponges, clams, soft corals and crustaceans. They act as nurseries for fish. Importantly, coral absorbs carbon dioxide from the atmosphere via the CO_2 being dissolved in the water. In 2016 an underwater survey of the Great Barrier Reef showed that 35% was dead or dying. If the temperature changes for too long, the coral expels the colourful algae, which it uses for energy. This leaves a bleached white skeleton behind that is then taken over by seaweed.

There are many threats to the coral reefs, including over-fishing, physical damage from boats and divers, and chemical damage from pollution. Some unsustainable fishing practices include using cyanide to disorientate the fish, which also poisons the smaller organisms on the reef. Other fishermen rip apart reefs with crowbars or use blast fishing that involves dynamite; such methods cause extensive reef destruction.

Nutrient cycling

The upper sunlit layer has a boundary called the thermocline, which keeps it apart from the dark, colder and calmer twilight zone below. The photosynthesising plankton do their work in the light layer but will sink slowly into the abyss when they die. Death means that all the nutrients they contain are consigned to the depths: the same is true for all marine animals, ranging from zooplankton to whales. All their faecal matter sinks too. As a result, all the nutrients in the upper layer are gradually removed and sent to the bottom. To maintain the food web, nutrients need to be replaced or else there will be a net loss.

Marine snow is not quite as pleasant as it sounds. It is made from faecal matter, sand, soot and other inorganic dust, dead bacteria and plankton. It looks like white flakes and provides food and nutrients to the creatures below. That which isn't eaten forms a thick ooze on the ocean floor. So, is this part of a food web or more of a unidirectional food chain, with the movement of nutrients to the bottom?

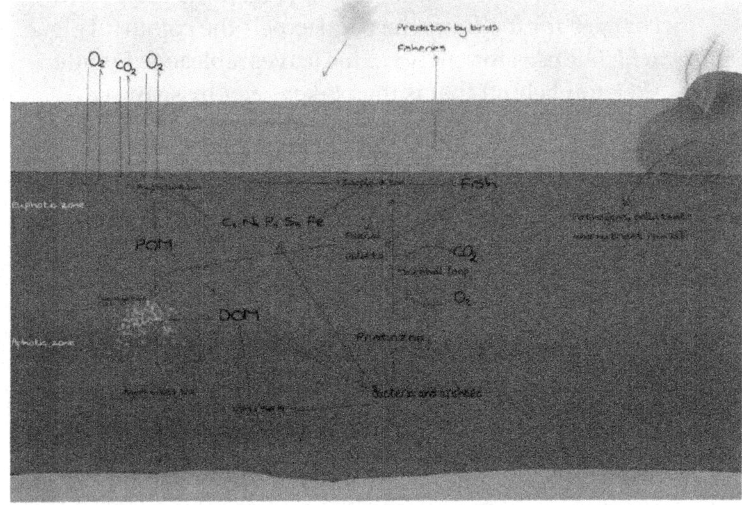

Fig. 8 Nutrient cycling: POM = photosynthetic material; DOM = dissolved organic material

The right balance of different nutrients is vital for the phytoplankton to grow. Eutrophication – an over-rich nutrient supply causing excessive plant growth - causes changes in the structure and function of the marine ecosystems. We have seen an increase in phytoplankton, algae and bacterial biomass; different species of algae thriving, which may be toxic or inedible to marine organisms; development of rooted macroalgae such as seaweed along the shores; fall in oxygen levels in deeper waters; high levels of death among fish and shellfish; and in coastal areas, coral mortality.

The deep water of the Pacific Ocean has a higher nutrient concentration than the Atlantic Ocean because it is older and so has had more time to accumulate nutrients.

Nutrients are more abundant in coastal and upwelling regions.

Carbon

Carbon is dissolved in the surface layers as carbon dioxide. Organic carbon is created by photosynthesis and strips this carbon from the upper layers of the ocean. Most of the carbon falling through to the midnight zone is used by other organisms; less than 1% forms a sediment. This is known as the biological pump; without it atmospheric carbon dioxide would be much higher. Carbon is also removed by the production and sinking of calcium carbonate or $CaCO_3$. Carbon dioxide (CO_2) from the atmosphere is combined with Ca^{2+} from river waters, and is balanced by $2HCO^{3-}$. For each mole of $CaCO_3$ that's formed, one mole of CO_2 is carried to the deep sea, and one mole of CO_2 is left behind in surface waters.

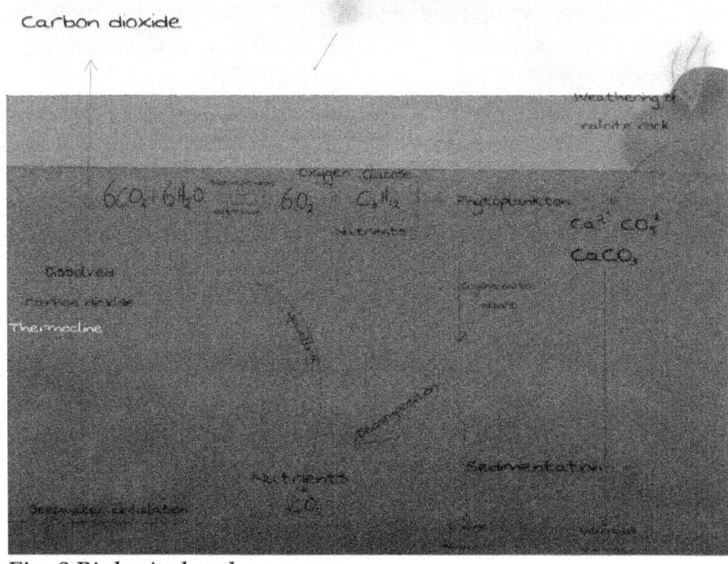

Fig. 9 Biological carbon pump

Nitrogen

We have mentioned how algal blooms causes low oxygen levels and result in the death of fish and overall decreased productivity. This causes changes to the availability of nitrogen in microbial communities. These nitrogen-rich phytoplankton sink to the sea floor when they die. They consume more oxygen as they decompose to create 'dead zones'. At low oxygen concentrations, most marine life is unable to survive. There has been a doubling of these zones every decade as more nitrogen enters the oceans from fertilisers used in agriculture and nitrogen-containing sewage effluent.

Nitrogen can be fixed by microorganisms within the oceans, but this process is enhanced when iron and phosphorus are present. More nitrogen run-off from agriculture stops diatoms (tiny marine phytoplankton) from using NO_3 as effectively. In San Francisco Bay there has been a long-term decline in productivity, attributed to higher amounts of nitrogen as NH_4. Humans add nitrogen via sewage effluents and habitat disturbances such as landfill, mangrove or seagrass clearance, and disruption of estuaries.

Iron

Whereas nitrogen places limits on the production and growth of plankton, low levels of iron in oceans can also limit the rate of growth. Since the early 1970s, the dust load for an expanding Sahara has increased nearly fourfold, and it has been shown that about half the dust reaching the Equatorial Atlantic is due to disturbed soil conditions. This dust contains more iron than undisturbed desert dust. As a result, there is likely to be more nitrogen available, which in turn is likely to increase the amount of carbon fixed.

Silicon

Silicon is an essential nutrient for diatoms as they form an important constituent of the cell wall. More diatoms means a higher take-up of silicon dissolved in oceanic surface waters, but the concentration of silicon increases below the photic zone after their death. If silicon concentrations fall other algal groups are able to outcompete diatoms, potentially leading to a toxic bloom.

The nutrient cycle works best if undisturbed by humankind. Removing top predators such as big fish, sharks and whales has a knock-on effect for other sea creatures. Sperm whales and humpback whales feed in the depths but return to the surface to breathe and defecate. It was thought the movement of these and other large creatures such as seals and turtles helped the water to circulate from the depths to the surface and replenish the minerals and nutrients for the phytoplankton. As they move, the current or slipstream behind them carried the nutrients upwards along with small creatures such as zooplankton. This causes migration up and down through the water column. To restore the productivity of the ocean, we need to sustain the populations of large animals.

Threats to the marine ecosystem

There are multiple threats, including the effects of pollution, chemicals and sewage. Heavy metals such as mercury have been implicated in birth defects and cancer. Toxins accumulate as you go up the food chain. If each plankton has one part per billion of mercury, but a krill eats 1,000, then their concentration is the same as one in a million. If a cod eats 1,000 krill, their concentration is now 1,000 in a million. This is bioaccumulation. The American health department advises women against eating certain fish when pregnant to avoid contamination.

Plastics are not usually biodegradable. In the Pacific Ocean there is the 'Great Garbage Patch' and in the Atlantic Ocean the 'North Atlantic Garbage Patch'. These are areas of manmade debris consisting of small, suspended plastic particles floating in the oceans. They measure up to 8% of the size of the ocean, although the exact size is very difficult to calculate. Microbeads are small plastic beads found in some cosmetic products. The plastics are unavoidably consumed by sea life, which seems to be affecting their growth and fertility and contributing to the collapse of some populations.

Closer to shore, pesticide and fertiliser run-off contribute to algal blooms that are in turn toxic and choking to other marine life.

As oil is transported around the world there is the risk of oil spills. Depending on the location of the spill, different creatures are affected. If close to shore, sea birds' feathers become clogged and death comes from exposure, drowning or starvation. Seashore plants are affected. Some bacteria can biodegrade oil and make use of the chemicals to produce energy for themselves. This leads to an overgrowth of certain species of bacteria, which upsets ecological balance and can cause further problems at a later date.

Over-extraction of marine life such as over-fishing is becoming unsustainable in parts of the world. 25-33% of marine species caught are fed to animals as feedstock. Overfishing causes accelerated growth rates of remaining fish, promotes earlier sexual maturity, and impairs the regulation of primary consumer populations, as has been shown in several species including Peruvian anchovy, Alaskan Pollack and North Sea cod. There are several options to overcome over-fishing such as changing food stock for animals, limiting catches, imposing bans in areas such as spawning zones, and changing practices to encourage selective catching with a rod and line rather than drag netting. There is damage caused by catching the wrong species and subsequent dumping of dead fish, while the seabeds are damaged by deep-sea trawling methods.

Summary

The ocean can easily be overlooked due to its vast size and the damage being unseen or forgotten. The marine ecosystem is complex, with multiple abiotic factors such as light levels, oxygen, carbon dioxide, temperature, pressure and nutrient levels all influencing life. We continue to see amazing diversity in extreme conditions. Continued exploitation and the removal and disruption to food networks may increase the number of 'dead zones' where life can no longer survive.

Chapter 13 - The woodland ecosystem

The diligent farmer plants trees, of which he himself will never see the fruit.

- Cicero

Examining woodlands gives us an opportunity to look at the interaction of other ecological factors and the impact on biodiversity.

Types of woodland

Woodland can be categorised into three main types: ancient woodland, semi-natural ancient woods, and pure plantations.

Ancient woodland

Woodlands first colonised the British Isles after the last Ice Age about 11,500 years ago; some ancient woodlands date back that far. By about 1000BC, the Iron Age was in full swing and about half the original 'wildwoods' had been felled for timber and for agricultural land. Detailed maps have been available since about 1600AD; as a result, the UK defines an ancient woodland as one that has had continuous tree cover since then. Generally, the older the wood the greater the biodiversity. Currently, less than 2% of the woodland in the UK is defined as ancient. In ancient woodlands the soils have been undisturbed and this has an impact on the plants and animals that have thrived. A mix of plants such as wild garlic, dogs mercury, yellow pimpernel and slow-growing fungi can be found and provide one of the richest wildlife habitats. A lot of resources are directed at trying to preserve and maintain the diversity of ancient woodlands.

Semi-natural ancient woods
These have developed naturally, whereas plantations on ancient woodland sites involve adding in new non-native trees, especially conifers such as spruce and pine, among the ancient trees.

Pure plantations
These often have only one or two species and are planted purely for harvesting of the wood on a cyclical basis. Competing plants are intentionally removed to allow the crop trees such as larch or pine to grow. Sometimes the trees are planted so closely together that a dense, dark shade pervades with little other growth occurring. Trees are later thinned to encourage more growth from the remaining trees. Conifers are able to grow up to six times faster than broadleaf trees and the softwood is used for everything from the production of paper to furniture-making. These tend to be even-aged, producing trees of the same height with the lack of variety afforded by the layering we will see shortly. We saw in Chapter 10 the problems with monoculture crops and the reduction in diversity of other forms of life that comes with it.

Niches

An ecological niche is the position in the environment that a species occupies. The niche will provide the food and shelter it needs and encompasses all the abiotic factors and biotic interactions and relationships.

A fundamental niche all the possible environments a species could live whereas a realised niche is where it actually lives. Some species cannot live in their entire fundamental niche due to competitive exclusion. Competition for resources, nutrients and water means one species may dominate over another. Competition can be asymmetrical or symmetrical but both influences species distribution. For example, if both jay birds and squirrels both compete for hazel nuts in a wood, only one of them can actually eat them. This may force the other species to a different part of the wood or to use a different food source. Shelter may be fought over by two species of birds wanting to occupy the same trees in the canopy layer. The competitive exclusion principle is that only one species can occupy the niche at any one time. Looking back to archaea from chapter 9, most live in their realised niche rather than their fundamental niche because in many environments they are out competed by bacteria for resources.

The structure of the bark of a tree is a niche in which different beetles will occupy different tree types. Bark acidity plays a role particularly in lichen growth. Different species need different amounts of light or combinations of nutrients or other abiotic factors.

Within a wood there are different layers seen in Fig. 1 and these have an impact on the niches available and the biodiversity present.

- Field layer with mosses, lichens and fungi
- Shrub layer with ferns and grasses
- Understory of young and low growing trees
- Canopy layer of mature trees
- Dead wood layers

Having a layered woodland improves biodiversity as each of these layers offers different habitats and niches for different plants and animals. There are not always all the layers in every wood. Densely packed trees (especially coniferous or evergreen trees) block light from reaching the field or shrub layer and it may be barren. The layers are not distinct and often blend into one another. This is the vertical structure of the woodland. Structural diversity leads to each layer providing different microhabitats for specific species.

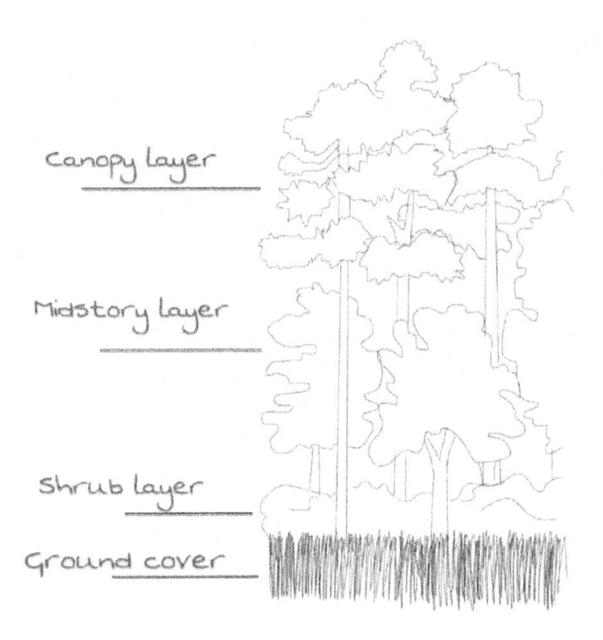

Fig. 1 Woodland layers

Rides or tracks cut through the wood along with open glades are also structurally important. Wildlife and plant life benefit from the extra light available and changes to the microclimate. Many species favour these woodland edges, where the land transitions from one type to another.

Some trees may be relatively few in number in an ecosystem, but may be the sole food source for some dependent wildlife. It is important to protect and maintain the variety of trees present. If one type of tree or plant has a poor year and produces little food or seeds, this can be offset for some consumers by finding food from alternative species. For example, a poor beech nut harvest may mean the squirrel eats more acorns instead.

The diversity of tree species is not the same as the biomass of the species. For example, a tree may only support a limited number of insects, but if they are present in large numbers, they provide a valuable food source. A single oak tree with thousands of caterpillars munching on its leaves will be favoured by any birds looking for an easy meal. Some species, such as rhododendron, support very little or no insect life. Rhododendron can out-compete native trees for light and resources. They provide a dense shade that affects other plant and animal life. They produce toxic chemicals known as 'free' phenols, which accumulate in their tissue. This protects the rhododendron bush against herbivore and insect attack and, without a natural predator or grazer, the rhododendron thrives. Honey made from rhododendron flowers causes 'mad honey disease'. It contains the toxin grayanotoxin, which can cause hallucinations in humans. A tale is told where, in 67BC, Roman soldiers fed on tainted honeycomb and were slaughtered due to their intoxicated, hallucinogenic state. Rhododendron can become the dominant species in an ecosystem through rapid growth and lack of predation and, by causing the destruction of other habitats, removes whole populations of insect, plant and animal life.

Interspecific competition

This is when two species compete for the same resource - in this case, rhododendron and a tree population such as birch trees.

Intraspecific competition
This is when two members of the same species are competing for limited resources - for example, two rhododendron bushes growing in the same place.

Resource partitioning
This can occur instead of one species out-competing another. Here, both end up evolving slightly and changing the resources they can use. An obvious example was seen with the finches on the Galapagos in Chapter 5, where the change in beak size and shape allowed different foods to be consumed. The same happens with woodland species and can lead to species co-evolving. Bees can have different length proboscises, which allow them to feed on different flowers. Over time, both the bees and the flowers diversify into two separate species occupying different niches. Each niche then has a bee species and a plant species that are co-dependent.

Keystone species
A keystone species is one upon which the ecosystem relies and it exerts a strong control. They may not be the most numerous or the biggest nor the most dominant, but they are a vital species. If the keystone species are removed, the whole ecosystem can collapse. One example is ground beetles. These beetles eat a variety of aphids and slugs and act as a pest control. They munch on the seeds of some weed species of plant. In turn, they are fed upon by birds and small mammals. Their numbers have fallen by up to 70% in some habitats over the last 15 years, although the populations were most stable in woodlands. If they are removed from the ecosystem the other species that rely upon them suffer a fall in their numbers.

Maintaining diversity with hedgerows

Hedgerows are vital in maintaining biodiversity. They are the main habitat of 50 species in the UK, including 13 that are threatened or rapidly declining. Hedges are important for insects such as butterflies and moths but also for birds and mammals such as dormice and bats. More than 600 plant species, 1,500 insects, 65 birds and 20 mammals have been recorded as living in or feeding on hedgerows.

Since 1945, there has been a dramatic loss of hedgerows in the UK. An estimated 50% loss in overall hedgerow length has occurred, declining from 662,000km to 328,000km. Some of the loss has been from destruction, mainly due to changing farming practices such as having larger fields that need larger machinery. Some loss is due to neglect, as traditional skills of cutting and laying of hedges decline. Other factors leading to loss include lack of replacement of old hedgerow trees, changes in pesticide and herbicide use, increased 'ranching' or netting of fields, and careless stewardship.

Collectively, lost hedgerows lead to leaching of the soils, changes to nutrient cycles, erosion, lack of water retention and worsening fertility leading to even more losses. Hedgerows effectively regulate wind speed across fields, leading to less water loss. They also improve water quality by filtering out some pollutants from reaching watercourses. Trees and hedgerows play an important role in reducing the risk of flooding after heavy rain. A return to more plants, leading to more insects, leading to more birds and higher-level consumers leads to a more balanced ecosystem.

Culturally, hedgerows mark our heritage of complex farming patterns over the millennia. Some hedges are more than 1,000 years old. They provide an aesthetically pleasing aspect to the countryside, adding pattern and texture to the landscape.

A thicker hedge provides more protection for wildlife. Nesting birds prefer thick bases, which also provide great opportunities for more insects to thrive. More tree and plant species lead to a more diverse group of foods being available, which attracts a more diverse group of feeding. More layers lead to more microclimates and more niches for different species. Farmers are required to have a one-metre-wide strip between the crop edge and the hedge. It is estimated that for every year left uncut a hedgerow will gain two new species of bird. Overall, a hedge of four metres in height suits the majority of wildlife, although some birds, notably yellowhammers, whitethroats and partridges, prefer a shorter hedge. Hedge trimming is done outside of the nesting season, but if it is done too frequently and too harshly there is a reduction in the amount of seeds and fruits that are produced.

Within hedgerows, biodiversity can be maintained by planting appropriate species, active management, keeping pesticides away from the bases, less frequent trimming and wider strips before the crop edge.

Active woodland management

Woodland management depends on the role and function of the wood. For some, the trees are used as a timber crop; for others, maintaining the biodiversity is more important. Removing non-native or competing species, controlling pests and predators and ensuring adequate nutrition and water all take active management. Having small areas of woodland that are not joined together means that species of plant and animal cannot travel and colonise new areas easily - especially plants and trees that disperse their seeds poorly and insects and animals that don't or can't travel long distances.

Coppicing is a skill that has been practised for many hundreds of years. It involves cutting trees on a regular rotation - usually between every seven and 25 years - to allow more light to reach the ground and resulting in an explosion of plant life. Additional light provides warming of the soil and stimulates seeds to sprout. Trees such as hazel, oak, maple, lime and ash are ideal for coppicing. Growth of the new stems can be rapid - in some cases as much as 5cm in just one day. An oak stem can grow 2m in a season, while some species of willow can manage up to 4m.

Coppicing and thinning allow improved nesting sites for birds, better plant growth due to less shading, and more insect life such as caterpillars and subsequently butterflies. The traditional skill of coppicing is a time-consuming and labour-intensive job; as it declines, there is a knock-on effect and decline in the numbers of insects and butterflies. The cut stems can be used as poles or rods for fence-making or furniture-making, fuels, tool handles, baskets and, historically, for ships' planking. Unfortunately, these products are of such low financial value that coppicing tends to be uneconomical.

Pollarding is similar to coppicing, but rather than the trees being pruned low down near to the ground, it takes place at a height of over 2m. It is often done where problems with animal grazing exist, as the tender new shoots are a temptation to a hungry herbivore.

Branching out into trees

English oak trees (*Quercus robur*) are probably the best-known and best-loved native British tree. They can grow up to 40m tall but only start producing acorns once they are over 40 years old. The acorns are a rich food source for many creatures, including mice, squirrels and birds. Their open canopy allows shafts of sunlight to reach the woodland floor, allowing plants to thrive below. A single oak tree can support well in excess of 300 species of insect on its own, including stag beetles and caterpillars. In addition, there can be 300 species of moth alone found on oak trees. However, not all insects treat the oak kindly. The oak processionary moth causes great damage - not just to the oak tree but also to humans. Their tiny hairs carry a toxin that can cause itchy skin, sore throats and breathing problems. Their caterpillars feed on the leaves, weakening the tree and making it more susceptible to other diseases. Acute oak decline causes serious problems to oak trees including death. The cause is currently unknown, although abiotic factors or a bacterial infection from a bark beetle may be to blame.

Beech trees (genus *Fagus*) often cast a very dense shade, and pure beech woodlands can often have a poorly developed field or shrub layer. Their leaves rot slowly and often forms thick layers on the woodland floor. Despite the lack of plants below, beech trees play host to more than 100 invertebrates and 200 species of lichen. Some fungi, such as the porcelain fungus, are associated with beech woodlands. Autumnal beech nuts provide a treasured food source for badgers, squirrels and birds. Squirrels, however, harm beech trees by stripping the bark. In severe cases where the trunks are 'ring-barked' (have complete removal around the trunk), the tree dies due to problems with water and nutrient flow.

Birch trees (genus *Betula*) are known as pioneer species: their seeds are light and wind-blown, so they are usually one of the first trees to colonise a new site. Birch is not long-lived - up to 80 years maximum - but provides shelter from the wind for other, often slower-growing, trees while they take root. They have a naturally high biodiversity of insects, with more than 300 species recorded.

Willow tree (genus *Salix*) bark has been used for well over 2,500 years as a remedy for fever and aches. The active chemical salicin is extracted and is metabolised into salicylic acid in the body. This is the precursor of aspirin.

There is a direct correlation between number of tree species in an area and the number of insect species. Some species of trees are home to more insects than others; for example, the willow (genus *Salix*) plays host to about 100 times more insects than the maple (genus *Acer*).

The importance of dead wood

Dead trees provide an important place for insects and birds that nest in holes to thrive. Up to 20% of woodland fauna is found in dead, dying and decaying trees, while 30% of birds nest in dead wood.

Fungi are decomposers extraordinaire. Lignin is the most abundant organic polymer on Earth and is responsible for the strength of trees. Cellulose provides a similar function in plant cells. Lignin is an important sequester of carbon from the atmosphere. With lignin, trees are able to grow tall and compete for sunlight as lignin helps transport water within the living plant itself. Fungi use one of several enzymes such as lignin peroxidase to break lignin down and recycle the nutrients back into the surrounding habitat.

Standing dead wood provides different niches from rotting wood on the ground. A threatened species of stag beetle lays its eggs inside dead wood and allows its larvae to develop. Holes in the trunks of standing trees provide nesting opportunities for woodpeckers and bats. Lying wood - whether wet or dry through and ranging from heavily shaded or not - encourages different species; more than 1,700 invertebrate species have been found to be present. The fungi growing on rotting wood provide specific microhabitats for species of flies, beetles and small moths. Different species of tree decay at different rates; oak trees take a notoriously long time, but provide a long-term continuous habitat for insect larvae, beetles and fungi.

Fungi

Fungi are vital to woodlands and, as they don't photosynthesise, the low light levels in some woods are unimportant. One role they play is the breakdown of dead plants and wood. We have seen in Chapter 11 that fungi often form a symbiotic relationship with other plants and organisms: for example, fly agaric forms an association with birch trees and chanterelle works with birch and pine. These relationships are especially important in nutrient-poor soils. The mycelium network allows trees greater access to water and minerals. The greater the species richness of trees, the more fungi types are present. Larger woods tend to have more habitat niches, which also mean a richer mix.

Some fungi live in the soil, others are associated with plant roots, the wood itself or the leaf litter. Each niche is occupied by different groups of fungi. Beautiful-sounding names such as penny bun (*Boletus chrysenteron*), blue spot knight (*Tricholoma columbetta*) and purple stocking webcap (*Cortinarius pseudosalor*) live with the plant roots. Jelly ear (*Auricularia auricula-judae*), stinkhorn (*Phallus impudicus*) and yellow brain fungus (*Tremella mesenterica*) live on the wood, whereas collared parachute (*Marasmius rotula*), russet toughshank (*Collybia dryophila*) and clouded funnel (*Clitocybe nebularis)* inhabit the litter layer.

Some fungi are pathogenic, such as *Ophiostoma ulmi*, which causes Dutch elm disease. The elm bark beetle is the insect vector. In the 1990s, more than 25 million elm trees were lost in the UK due to the disease.

Lichen

Lichens are a fungus and an alga working together in a symbiotic relationship. Some lichens are more than 8,000 years old, and there are around 1,700 species in the UK. The three main types are crustose (crusty), foliose (leafy) and fruticose (shrubby).

One leafy lichen is the tree lungwort, as shown in Fig. 2. It is used as an indicator species for ancient woodlands. The fungus is *Lobaria pulmonaria* and the photosynthetic algae is *Dictyochloropis reticulate.* If it weren't for the fungus, the algae could not survive out of water. Equally, the fungus needs the algae to provide the energy by converting sunlight. To make the relationship more interesting, there is actually a third party, the cyanobacteria *Nostoc sp.*, which fixes nitrogen from the atmosphere. Tree lungwort is sensitive to pollution and has been in decline across Europe. Losing tree lungwort leads to a significant loss of nutrients as it fixes large amounts of nitrogen. Interestingly, it has been used as a flavouring in Siberian beer.

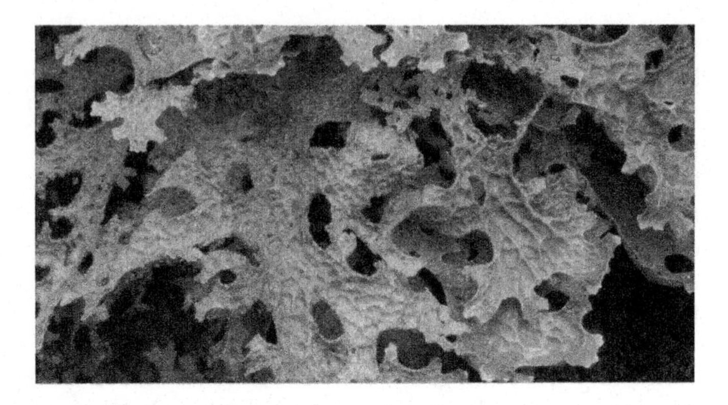

Fig 2. Tree lungwort

Plants

'Indicator plants' are used as a guide to the soil type present. Rich, dry soils have burdock, bluebells, hazel, ivy and elder, whereas rich, moist soils have nettles, ground ivy, tufted hair-grass and dogs mercury. Very wet, rich soils show growth of angelica, meadowsweet valerian or golden saxifrage. Poor dry soils are more likely to have cowberry, wavy hair-grass and common bent. As the moisture content rises, heather, bilberry, wood sorrel and purple moor-grass predominate. The plants present can reflect the pH level; for example, heather hates lime-rich soils and poor drainage.

Nettles are a very useful plant. They grow in fertile soils and provide an early food source for bees before other flowering plants are in bloom.

Woods have different combinations of plants depending on where in the country they are situated and the species of trees present. Plants that thrive in low light levels are commonly seen and often flower early in the year before the leaves appear on the trees. Primrose, bluebell, dog violet and wild garlic all flower in the springtime. The age of a woodland can be surmised using the Ancient Woodland Indicator Species guide. These plants tend to be slow-growing and only colonise woodlands gradually. Examples of the plants include opposite leaved golden-saxifrage, wood sorrel, wild daffodil and wood anemone. Bluebells are perhaps the flower most commonly looked for, but there are actually three distinct types: English, Spanish, and a hybrid of the two.

Animals

There are a great many types of insects that thrive in woodlands, with many finding specific niches. They are too numerous to list here, but can range from froghoppers and leaf hoppers that feed on plant sap; moth larvae that feed on leaves; stem nesting wasps that use bramble; and purple emperor butterflies that feed on aphid honeydew. Some animals both large and small can cause damage to trees, from bacteria and fungi, to squirrels stripping beech tree bark, and deer and rabbits damaging young trees through grazing. Grey squirrels carry the squirrelpox virus, which is fatal to red squirrels.

Uses of woodlands

Woodlands have had many uses over the millennia. Originally they were used for agriculture and food production. Hazel trees produce hazel nuts as a valuable food source for humans, birds and animals such as squirrels or mice. Currently, the Turkish are the greatest hazelnut producers, growing 75% of the world's crop.

Maintenance of pollinators for other crops is vital; for example, the willow tree produces nectar for bees and is a valuable early pollen source.

Culturally, woodlands provide an escape from urban life and allow for recreation, improvements in health and well-being and are used for education purposes.

Wood products have been used in basket-making, fencing, hand tools, walking sticks, bowls, spoons, bows and furniture-making.

Drugs have been discovered such as aspirin from the bark of the willow and quinine from cinchona bark.

Fast-growing trees such as willow or hazel can be grown as a biofuel crop. Some countries such as Sweden harvest and burn the crops on a large industrial scale. In many parts of the world, wood is the main fuel for cooking. Historically, charcoal was used in iron smelting, bark from oak used in animal hide tanning, and ash from burnt wood used to produce potash for glass- and soap-making.

Wooded areas provide noise reduction, flood regulation and soil protection, and are a treasured feature of the landscape. Trees such as willow are environmentally important as they are used in water bio-filtration and wastewater-management systems. Trees help with land and soil reclamation, slope stabilisation and flood prevention schemes. Flooding in Pickering in North Yorkshire between 1999 and 2007 caused an estimated £2 billion worth of damage. Previous wisdom was to spend money on a concrete wall to protect the town's shops and houses. However, planting 20 hectares of woodland, using logs to stagger flow in the local becks, and building an embankment provided protection at a tenth of the cost of a concrete wall. With climate change predicted to increase the number of severe weather events, tree planting may benefit local communities and ameliorate the problem. Tree planting has improved the number of wildlife habitats far more than a concrete wall would have done.

Threats to woodlands

Threats exist to woodlands themselves and to the countryside by the removal of the woods.

The biggest threat is the continued encroachment of humankind in the development of housing and transport links. Changes to farming and agricultural practices means the loss of hedgerows and neglect of woodlands. Loss of traditional skills such as coppicing, hedge laying and woodland management continue. Apathy towards ancient woodlands will only accelerate the declining proportion of wooded areas in the country.

Individual species of tree are under threat. Ash dieback was seen in Chapter 10. The Living Ash Project was launched in 2013 and aims to identify which trees resist ash dieback and screen them using genetic markers. When the seeds are secured for breeding purposes the resistant trees can be replanted. If healthy ash trees are seen they can be reported at livingashproject.org.uk. Elms were badly affected by Dutch elm disease and oaks are under threat from the oak processionary moth.

The denuding of the countryside has removed one of nature's sponges to prevent floods, retain water and stabilise the soils. Trees enable rainwater to penetrate the ground 60 times faster than grassland.

Climate change will affect the balance of species that will exist in the woodlands of the future. The niches occupied by various trees, plants and animals today will alter. Some species will gain from a higher temperature and higher carbon dioxide levels, but others will lose out. The changing climate will transform weather patterns including amounts of sunshine and rainfall. Growth rates may accelerate for some in lengthening growing seasons as temperatures rise. Pollination rates due to shifting levels of different pollinators will affect the spread of plants and trees. Patterns of disease will alter, with some pests, such as the Asian longhorn beetle, moving to warmer climes.

Summary

Using woodland as an example, different aspects of ecology can be seen. The wide variety and diverse nature of the bacteria, fungi, plants and animals that coexist need to be maintained to continue to have a thriving woodland ecosystem. The benefits of investing in and maintaining woodlands can clearly be seen.

Chapter 14 - The future of genes and diversity

Genetic engineering is to traditional crossbreeding what the nuclear bomb was to the sword

– Andrew Kimbrell

There have been many scientific advances since the discovery of the structure of DNA by Watson and Crick (along with Rosalind Franklin and Maurice Wilkins) in the 1950s. Manipulation of DNA has become an increasing area of research, but there are benefits and hazards and many social and ethical implications to be considered.

There have been a large number of important microbial-based products and applications, which have already been developed. A few more will be examined in this chapter.

What is genetic modification?

Before we think about the latest techniques, we need to wind the clock back to when Mendel discovered the genes of pea plant breeding seen in Chapter 5. This explained the basis behind the breeding of 'better' plants and animals that had been occurring since the beginnings of agriculture.

Selective breeding

Selective breeding is selecting plants or animals with specific qualities with the aim of producing offspring with the same or enhanced qualities. Thoroughbred horse breeding takes fast or successful racehorses and breeds them with mares to have successful offspring. Over several generations, breeders end up manipulating traits within the species such as its size, growth rate or speed. While breeding can be successful, there is a lot of chance and luck involved. There can end up being a small gene pool with a lot of inbreeding, which can lead to genetic weaknesses being passed on. Overbreeding produces more horses than are needed, which means 'waste' and 'surplus' horses are slaughtered for their meat.

Selective breeding can be applied to other areas of farming whether it be cattle, chickens or wheat crops.

Crossing

Crossing is similar to breeding and involves taking two similar organisms and mixing them. Peas or tomatoes would be typical plants. The first generation cross is when two varieties that have desirable characteristics are crossed together. If the subsequent generations don't carry the desirable qualities of the first generation, you can go back to the original plants and cross them again.

Genetically modified organisms (GMOs)

So far we have used a plant or animal species and bred two together and waited to see what traits the new offspring hold. This doesn't sound like genetic modification but it really is. More commonly people tend to think of GMOs when it has involved the transference of a gene from one organism and placing it in the DNA of another species. This is called 'transgene addition'. These techniques have many uses: for example, increasing the resistance of a plant to certain pests, or adding nutrients to a food that they don't naturally possess, or enabling a plant to grow somewhere that it wouldn't naturally thrive. These techniques target specific genes for desired qualities in specific plants or animals.

Crop yields, taste of food, speed of growth or improved production in harsh conditions are all traits farmers want for their produce. Selective breeding or crossing takes a long time to see results, so directly manipulating the genome provides faster results.

GMOs that have had the transgene added are controversial mainly because the long-term effects of this type genetic manipulation are not known. Will the gene stay in the original target species or could it jump to another species and enter wild plants or animals? A survey in 2016 showed that 88% of scientists from the American Association for the Advancement of Science thought GMO food was safe to eat, but only 37% of the public agreed. Some of this is due to fear of the unknown, or disgust at the idea, or concern that big businesses have ulterior motives.

Papaya is an important food in Hawaii and the developing world but suffered crop destruction from a disease called ringspot virus. A gene for the viral coat was transferred into the papaya plant and provided immunity. This saved the papaya plantations, but there was a campaign to stop Hawaiian papaya from entering the food chain and this campaign was as damaging as the original infection. Papaya grown prior to the gene modification had *entire* viral particles on the fruit and leaves and had been eaten by humans for many years with no ill effect.

Modifying and cutting up DNA

Before we can add a gene, first we need to cut it out of the original DNA. We then need to check and examine it to make sure it is the gene we are after and then make copies to insert into our chosen DNA.

We want a specific gene. Rather than replicating an entire genome, we need to break DNA into smaller segments and can do this using an enzyme called a restriction endonuclease. These enzymes can cut or break DNA at very specific points by matching themselves up with a string of base pairs, binding to and then cleaving the DNA. For example, if the gene we were interested in started CTTGTC then we know it would bind to a mirror image of bases - i.e. GAACAG. If we manufacture GAACAG and add it to a restriction endonuclease it can only bind to the desired segment of DNA and a very specific cut is made.

One technique is zinc finger nuclease, another is TALEN, and the latest is CRISPR Cas9.

Zinc finger nuclease is a technique using a zinc finger transcription factor combined with a restriction endonuclease. A zinc finger is a small protein wrapped around a zinc ion, which, when combined with a DNA binding loop, looks a bit like fingers. A transcription factor is a protein that helps with the transcribing of the DNA and so preferentially binds to DNA and RNA. These nucleases were first discovered in the African clawed frog (*Xenopus laevis*). The restriction enzyme has at least 24 base pairs and is stitched onto the zinc finger to give a very specific cut. Two 'sticky ends' are created and produce a short(ish!) strand of DNA that can be subjected to a technique known as PCR. This amplifies the amount of DNA available and is described below.

If these single genes are added to DNA from another species that has undergone the same process to create identical sticky ends, then some of the new DNA will be taken up by the new host.

TALEN stands for Transcription Activator-Like Effector Nucleases; these are artificially created endonucleases. They were discovered in a plant pathogen called *Xanthomonas*. They consist of 34 amino acids that bind to specific nucleotide sequences; by adding an endonuclease, a cleaving occurs at a specific site. This technique has been used to alter plants, animal and fungi.

CRISPR Cas9 is the new kid on the block. It stands for Clustered Regularly Interspaced Short Palindromic Repeats and it is able to recognise specific sequences of DNA. Two scientists (who have not yet won a Nobel Prize, but may do one day!), Professors Jennifer Doudna and Emmanuelle Charpentier, used a nuclease from bacteria *Streptococcus pyogenes* called Cas9. Originally the function of the gene was to remove foreign or invading genetic material from the bacteria. By attaching the CRISPR and Cas9 together, very precise or specific cuts can be made using RNA as a guide. Then a piece of new DNA can be inserted by the cell's double-strand break repair machinery. This CRISPR-based technique can efficiently correct point mutations without cleaving DNA. This is likely to be at the forefront of genetics for many decades to come.

Polymerase chain reaction (PCR)

PCR assists in the analysis of the DNA fragments we have made by multiplying their numbers. This method was developed in 1984 using the thermophile bacteria *Thermus aquaticus,* which is found in hot springs. Normally, if DNA is heated to 95°C the two strands separate during the 'denaturing' stage. If 'primers' are added they will attach to their respective binding sites on each strand as the mix is cooled to 56°C in the 'annealing' stage. If there are the four nucleotides - A, C, G and T - around, the mixture heated to 72°C, and the DNA polymerase from *Thermus aquaticus* (called *Taq* polymerase) is added, then tens of thousands of base pairs can be added in just a few minutes during the 'extending' stage. The separated DNA strands join back together to form a double strand once again. Repeated cycles of heating and cooling magnify the initial stretch of DNA many thousands of times.

This is a fast, reliable and accurate automated process. All that is needed is: 1) The target DNA; 2) a DNA primer; 3) *Taq* polymerase; and 4) nucleotides. With these few ingredients, PCR can make a billion copies in just a few hours! PCR testing kits can be used to identify the parasite *P. falciparum*, which is responsible for malaria but uses a specific 18S rRNA gene common to all malarial parasites.

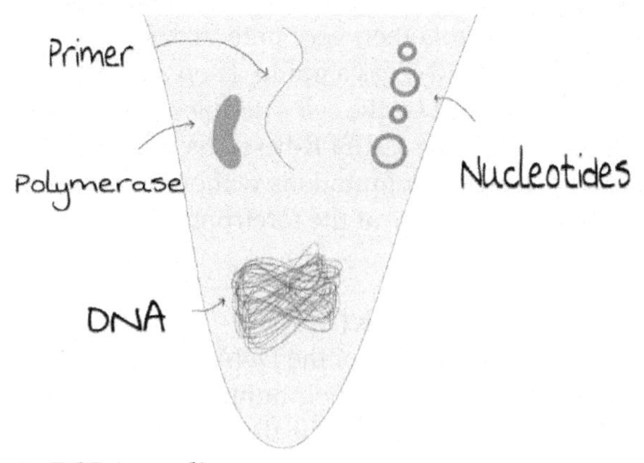

Fig 1. PCR ingredients

A reverse transcriptase can be added if RNA is the starting molecule rather than DNA.

Agarose gel electrophoresis is able to separate out DNA fragments according to their size. We know from Chapter

2 that because of the phosphate backbone, DNA has a negative charge. If we put a positive electrode at one end of the gel, negative molecules are attracted and move towards it. Agarose gel acts like treacle and slows larger fragments down more than the smaller fragments. This gives a series of bands in the gel based on size, as shown in Fig. 2, and are called 'ladders'. Genes for bioluminescence can be added that make the bands glow in the dark or for checking that the right cells have taken up the right genes.

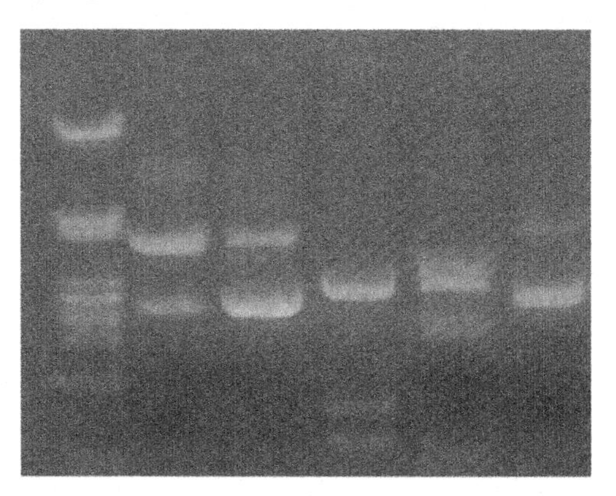

Fig. 2 Gel electrophoresis bands

Adding genes to DNA and having a sticky end
We can now cut the precise gene we want from DNA with a restriction endonuclease technique, multiply the numbers many million-fold using PCR and check we have the right one with agarose electrophoresis. Now to add the gene to our destination DNA.

We have seen how SIV and HIV retroviruses use reverse transcriptase enzymes to convert RNA back into DNA and insert it into the host's DNA. Genes can be added to DNA by using these retroviruses as a vector. However, in some cases the newly added gene could be inserted into a place on a chromosome where it was neither needed nor wanted - for example, in the middle of a functioning gene that causes that gene to stop working.

To be more precise, we need a way of targeting only the specific site of interest.

Restriction endonucleases that were initially discovered in bacteria such as *Hemophilus aegypticus* generated an enzyme named HaeIII. This enzyme specifically cuts a section with GGCC in it and gives a 'straight cut'. In our bodies, enzymes from the pancreas break down DNA in this way but at random rather than specific points. To add one gene into a stretch of DNA, we need a 'sticky end' or offset cut. To create a sticky end needs a particular sort of restriction endonuclease such as HinDIII or EcoRI from *Haemophilus influenza* and *E. coli* respectively.

DNA ligase is the enzyme we met in Chapter 2, which reforms the fragments of DNA permanently back into a long strand called recombinant DNA. Since transcription and translation mechanisms and the genetic code are universal and by combining the enzymes above, a technology is created that can transfer fragments of DNA from one organism or species to another.

Bacteria

Genetic techniques discussed above have allowed evolutionary relatedness to be more precise and detailed. We saw in Chapter 5 that using the phenotype of an organism created problems, for example, by linking all creatures with wings together when they have had different evolutionary pathways. Since DNA was discovered and genetic markers were found to be passed from generation to generation, scientists and evolutionists have been able to link organisms that were previously apparently unconnected. Being able to identify the genes for the ribosomes - specifically the small subunit and the 16S rRNA gene - has opened up new chapters in the discovery of the history of the planet.

Industrial uses of bacteria

New ways of using bacteria spring up daily across the scientific and industrial world.

The extremophiles we met in Chapter 7 were a vital discovery. Many industrial processes such as PCR work best at higher temperatures. Unfortunately, the proteins and enzymes can become misshapen and deformed when heated, in a process called denaturing. Denaturing happens to an egg white when it is cooked and is transformed from a clear jelly-like liquid to a solid white mass of changed protein strands due to the application of heat. The DNA polymerase from *Thermus aquaticus* is an example of a stable enzyme able to operate at higher temperatures.

Fuel and energy production

Biofuel production can be done by modifying or tweaking bacteria or fungi. Yeast is an organism that is able to create alcohol from sugar and water. Alcohol ends up being toxic to the yeast cells and they die off. The trick would be to produce a non-toxic fuel, remove the fuel before it reached high enough concentrations to cause damage, or modify the yeast to be resistant to the effects of the alcohol. The latter would be a biofuel-tolerant organism. The enzymes in some soil microorganisms can ferment sugars, starches or (less commonly) cellulose into bioethanol. Cellulose is a constituent of wood and other indigestible parts of plants. Scientists could rely upon evolution and keep exposing bacteria to higher concentrations of the fuels to select out the more resistant ones and allow these to grow and be cultured. However, taking a resistance gene from one strain or species and transplant it into the original species could fast-forward the otherwise slow steps of natural selection.

Unfortunately, there are many problems with producing biofuels. Biofuels may not be any cleaner than today's fossil fuels, as they are usually carbon-based and generate carbon dioxide when burned. Additionally, many biofuels are grown on the land we use to grow food crops. If less food is produced, causing food prices to rise and leading to food shortages, the poorer parts of the world would be disproportionally affected.

Algae show great potential in producing energy from sunlight. They are little chemical factories running on sunlight, water and carbon dioxide and producing oxygen and energy. This happens in nature already and, as algae are inexpensive and don't tend to process noxious waste chemicals, they could solve some of the global energy problems. If energy-rich strains of algae are identified, their genes can be analysed and these can be mixed into the genes of other algae or photosynthesising bacteria. These would generate more energy for the same amount of sunshine.

Humans

Gene technology has helped some and hindered others. At crime scenes, any DNA found is analysed and generates a 'genetic fingerprint'. This genetic fingerprinting uses the PCR method described above to analyse DNA fragments and compares the suspect's gene profile to the genetic relationships and genetic variability within the population. It looks at the variable number tandem repeats (VNTRs) or short tandem repeats (STRs). STRs are short sections of DNA between two and ten base pairs long that occur at specific non-coding points on a chromosome and are repeated many times. The number of repeats varies from one person to the next. Imagine if the base pairs CGCT were repeated five times at a specific location on a chromosome and the known frequency in the population was one in ten. If a victim had unknown DNA with the repeated STR, then 10% of the population would be potential suspects. However, if on another chromosome the code read TGGA, was repeated three times and was present in only 5% of the population, the likelihood of finding a person with both the first AND second variation would be 0.1 x 0.05 = 0.005, or 5 in 1,000. Combine this with more and more STRs and the likelihood of having another person with exactly the same combination of number of STRs becomes smaller and smaller. As the probability of two individuals having the same STRs reduces, the confidence of arresting the correct suspect increases. If 13 STRs are identical between the DNA on the victim and that of the suspect, what are the odds of it being someone else? About one in ten trillion.

The use of genetic fingerprinting has uses in the fields of forensic science, medical diagnosis, and animal and plant breeding. During the UK horse meat scandal in 2013, when horses entered the food chain and were labelled as beef, food fraud was detected by analysing the DNA of the products on sale.

DNA extraction at home
You will need:

- cold tap water
- teaspoon of salt
- washing-up liquid
- strong alcoholic spirits - preferably over 70% alcohol - in the freezer
- DNA from fruit such as strawberries, or wheat bran
- Coffee filter paper

Step 1 Break apart the plant matter. Squish up the fruit in a glass or plastic bag until it looks like a mushy mess.

Step 2 Break open the cell walls. Chapter 1 showed us the phospholipid bilayer; to break that up we need some soap, so add 100ml of cold tap water, 1/4 teaspoon of table salt and two drops of washing-up liquid. Gently mix, trying not to make bubbles, until the salt dissolves. The salt helps the DNA stick together. Leave the mixture to rest for 10-20 minutes; this allows the soap time to release more DNA.

Step 3 Filter out the plant matter. Carefully pour your fruit mixture through the filter and into a clean glass. You can gently squeeze the filter to get more liquid out. DNA is soluble in water and contained in the filtered part of the solution. Pour this into a test tube if you have one.

Step 4 Separate out the DNA. Using the ice-cold alcohol makes the DNA come out of solution. Tip the test tube to a 45-degree angle and very **carefully pour the alcohol down the side to create a layer on top of the filtered solution. Alcohol is less dense than water and so floats.**

DNA will appear as white threads in a layer between the water and alcohol as it precipitates out. This is a sticky mass of DNA molecules!

Diseases

Humans with single-gene disorders might have some of their cells first removed, then treated in vitro by zinc finger nuclease or CRISPR Cas9, and then returned to their body.

We are now a step closer to correcting single-gene diseases in humans with patient-specific cell transplants, while techniques such as CRISPR Cas9 offer hope for curing such diseases in the future.

We saw how muscular dystrophy was caused by abnormalities in the introns and exons of a gene in Chapter 3. Cystic fibrosis (CF) is another inherited genetic condition. It is caused by a recessive gene and, like Mendel's peas, needs two faulty copies to be present before the disease is present. It affects one in 3,000 births. Historically, mothers recognised a child with CF by 'having a kiss on their salty brow'. The gene is Cystic Fibrosis Transmembrane Conductance Regulator (CFTR) and codes for a protein that moves chloride ions in and out of cells. It is located on chromosome 7 and is made of a sequence of 250,000 nucleotide bases. If the gene is not being expressed properly, the mucus in the lungs and pancreas get thicker and has a glue-like stickiness, and the sweat changes to being extra salty.

The first diagnostic test developed was the 'sweat test', which measured the amount of chloride in the sweat. In the lungs, the sticky mucus traps more bacteria and cannot be removed by the natural mechanisms such as the cilia (little hairs lining the airways). Being a gene of 250,000 nucleotides, many different mutations can occur. More than 1,000 have been identified so far, but the most common involves a deletion of just three nucleotides. This loss of a codon means the loss of an amino acid. In the protein sequence a phenylalanine amino acid is no longer where it should be - at position 508. This is the case in 90% of cystic fibrosis sufferers. Other patients can have a missense mutation where aspartate is inserted instead of glycine or a shortened protein by coding for a stop codon. All of these mutations lead to the protein being folded up wrongly and as a result, there is a complete loss of ability to function.

Gene therapy for CF is being explored to get a normal copy of the CFRT gene into affected cells using different vectors such as liposomes or viruses. At the time of writing, no effective treatment has been found.

The Enzyme-Linked Immunosorbent Assay (ELISA) test checks for antibodies made in response to infection. A patient would produce specific antibodies if they have been exposed to a specific infection. The ELISA test uses antigens that will bond to these antibodies. A second antibody with a marker is added that binds to the patient's antibodies if present and changes colour. If the patient hasn't been exposed and therefore doesn't have any antibodies, then the second antibody would not have anything to bind to and hence there would be no colour change.

Food

Biodiversity is invaluable for biotechnological innovation. So far much microbial biodiversity is largely unexplored. The application of biotechnology in agriculture has resulted in new crop varieties with increased resistance to pests and diseases, as well as crops with increased nutritional value (such as golden rice).

However, changing varieties of staple crop is nothing new. A simple question would be to ask what colour carrots are. Most people would say orange, but during the 17th century purple, yellow or white would have been correct. The Dutch were known as carrot farmers and developed a strain of carrot that contained higher amounts of beta-carotene - the first orange carrot. The next easy question is; what colour was corn (*Zea mays* or maize) from a corn on the cob originally? Yellow? Not only yellow but red, white, blue, purple, pink and black too, and not only solid colours but spotted or striped.

Maintaining the pollination of crops is vital. In both the UK and America one can rent a bee - although not singularly! Honeybees that live in hives can be ferried about on trucks. They might be used in California for the pollination of almond trees, then go to Washington for apple trees, followed by Dakota for sunflowers, and so on.

Technology to ensure crops are suitably watered uses aircraft or aerial drones and cameras to measure and analyse different light levels. Different wavelengths of light reveal variable levels of pigments – green chlorophylls, orange carotenoids and red anthocyanins. The areas that show they are in need of food (or fertiliser) or water are targeted, sparing the crops that are adequately supplied.

Other species have contributed to changing foods in our diet. The bacteria *Xanthomonas campestris* causes some plant diseases but also produces a polysaccharide known as Xanthan gum. This gum is used in food production (as a thickening agent or in gluten-free bread); the oil industry (to thicken drilling mud or added to concrete poured underwater to help it thicken); cosmetics (to prevent separation of ingredients), the toy industry (in slime or fake blood); and in pharmaceuticals.

Yarrowia lipolytica is a yeast used in the production of cheese and other fermented foods. It releases enzymes that break down fats and proteins. The nutrients generated are useful both for itself and for neighbouring microbes. As it is so good at producing enzymes, it is often used in biotechnology. Genes are spliced into the DNA of the yeast to enable it to make and secrete specific proteins. It also produces and accumulates oil in its cells, which makes it a possibility as a biofuel.

Disease control

Viruses that attack bacteria are called bacteriophages. 'Phage' is another term for virus. Some of these are being developed against bacteria that cause food poisoning. In 2016, products against *Salmonella. E. coli* and *Listeria* were available. They are only effective against specific bacteria and unable to colonise human cells due to the differences in specific enzymes, structures and mechanisms of action. However, we have seen some of the defence mechanisms that bacteria can employ such as biofilms and have seen viruses crossing the species barrier such as with SIV and HIV.

Maintaining and improving future biodiversity

First, we need to recognise the importance of diversity and the fact that it is vital for the survival of the human race. By understanding the roles of different organisms in the complex food webs and networks, and appreciating the problems that upsetting the natural balance of prey and predator can result in, then the variety will be valued.

Second, we can tackle the problems around us - whether it be planting trees for flood defences rather than building concrete walls or being aware of the consequences of deforestation or microbeads in the waters - by looking for ways to maintain and improve diversity rather than reduce it.

Third, continuing our exploration and discovery of new species - whether it be in the soil beneath our feet or in far-flung corners of the world - means we can categorise and classify the organisms found. Understanding their roles and unique approaches, whether it be energy generation or waste production, gives us an insight into where valuable organisms thrive.

Seed banks are one way to secure genetic diversity of plants. The Crop Trust runs the Svalbard Global Seed Vault in Svalbard, Norway. Presently, more than 865,000 samples, out of an estimated global total of 1.4m plant varieties, are stored at -18°C. Each year the number of stored samples rises along with the estimated total number of species. Other international organisations, including Biodiversity International and the Millennium Seed Bank Project at the Royal Botanic Gardens, Kew, also store seeds. Kew's Millennium Seed Bank holds 94% of UK native plant species in its vaults. There are other national, regional and local stores in countries across the world. Some store their seed samples in liquid nitrogen at temperatures of –196°C. Different seed banks specialise in seeds from specific areas such as dryland areas or crops such as wheat or rice. Some seed banks have been lost due to power failures, fire and floods such as in Indonesia, or in war zones such as in Afghanistan, Egypt and Iraq.

The loss of biodiversity of plants is shown in the history of seed catalogues. In the 1900s, some offered more than 400 varieties of peas but now almost all of the pea crops in America are grown from just two varieties. In China, an estimated 90% of rice varieties have been lost. If agriculture relies upon only one or two varieties of a certain crop and these are decimated by a disease or pest, the ability to feed the world's population diminishes rapidly. We have seen this recently with ash and beech trees in the UK or in the 1970s in America with the wheat crop losing 25% of its yield due to a fungal infection.

Summary

There are exciting prospects on the horizon but also many challenges. The planet has some difficult decisions to make about genetic manipulations not only of plants and animals, but also of humans. Whether techniques will become the preserve of the rich and wealthy or whether people from all nations will be able to benefit is not yet known. As the population grows, more thought needs to be given to feeding and sheltering all people. As land is used for people or crops, less is available for the variety of organisms that make up the complex ecosystems that allow the planet to survive.

Further reading

The variety of disciplines involved from astrophysics to geology and biology to chemistry means a broad range of topics should be read and the jigsaw of knowledge pieced together. Journals such as Nature (who supplied the 'A new view of the tree of life' in Chapter 7 are valuable for cutting edge information (Author: Laura A. Hug et al; Publication: Nature Microbiology; Publisher: Nature Publishing Group; Date: Apr 11, 2016 - available at (www.nature.com/articles/nmicrobiol201648#f2)) and are well worth reading.

There are many great authors writing far better books than mine. There are historical texts from Charles Darwin to those by E. O. Wilson, Carl Woese and Lynn Margulis. Of current day cutting edge scientists Nick Lane is an award winning writer who has some stunningly good books.

There are many TED talks such as those on slime moulds which are fascinating (www.youtube.com/watch?v=2UxGrde1NDA and be a-maze-d!). Websites for the Rainforest Action Network (www.ran.org), the Green Party (www.greenparty.org.uk) or WWF (www.wwf.org.uk) are worth exploring among many others to help prevent the loss of biodiversity in the world.

The ultimate test of man's conscience
may be his willingness to sacrifice
something today for future generations
whose words of thanks will not be heard.

Gaylord Nelson, co-founder of Earth Day

Glossary

Abiotic	non-living factors in an ecosystem
Adenine	one of the four bases of DNA
ADP	partially charged energy storage molecule
amino acid	building block of proteins
AMP	molecule able to store energy
Antigen	a molecule capable of inducing a response from the immune system
Archaea	one of the three domains of life
ATP	fully charged energy storage molecule
ATP synthase	protein complex which stores energy as ATP
Autotroph	organism generating its own energy
Avogadro's constant	number of atoms or molecules in one mole of a substance
Bilayer	two layers e.g in cell membrane
Biotic	all living factors in an ecosystem
Capsid	protein shell of a virus

Cellulose	an organic compound forming plant cell walls
Chemolithotroph	organisms able to use inorganic compounds as a source of energy
Chloroplast	green plastid within a plant cell where photosynthesis occurs
Codon (triplet)	three DNA bases coding for an amino acid
Covalent bonds	strong chemical bond involving sharing of electrons between atoms
Cristae	ridge or crest formed by infolding of internal membrane of a mitochondria
Cyanobacteria	bacteria which obtains its energy via photosynthesis
Cyclical AMP	cellular messenger molecule
Cytochrome	iron containing protein involved in ATP generation
Cytoplasm	internal fluid inside a cell
Cytosine	one of the four bases of DNA
Diffusion	movement of molecules from region of high concentration to low concentration
Disulphide bond	covalent bond between two sulphur atoms in a

	molecule such as an amino acid
DNA	molecule carrying genetic instructions
DNA gyrase	enzyme which uncoils DNA
DNA helicase	enzyme which 'unzips' DNA into two strands
DNA ligase	enzyme which joins together DNA fragment
DNA polymerase	enzyme which adds bases to lengthen DNA
Endoplasmic reticulum	network of intracellular tubules involved in protein and lipid synthesis
Endosymbiosis	theory explaining transformation from prokaryotic cells into eukaryotic cells
Eukaryote	cell with internal organelles
Exon	coding section of DNA. Part of the gene that will be translated into a final protein (see intron)
Fatty acid	long chain of hydrocarbons and a carboxyl (COOH) group
Gini-Simpson index	means of calculating diversity

Gram negative/positive	test to discriminate between different type of bacteria by cell wall
GTP	molecule which can be used as an energy source or as a cellular messenger
Guanine	one of the four bases of DNA
Haber process	industrial process for fixing nitrogen
Heterotroph	organism which consumes others for energy
Histone	protein which packages DNA by acting as a spool
Hydrophilic	water loving
Hydrophobic	water hating
Hydroxyl (group)	molecule ending with a -OH group where one oxygen atom is bound to one hydrogen atom
Hyphae	branching filament of a fungus
Intron	'non-coding' section of DNA (see exon)
Ionic bond	bond between oppositely charged ions
Keeling curve	graph of carbon dioxide in the atmosphere as measured at the Mauna

	Loa Observatory, Hawaii
Kreb cycle	series of chemical reactions used by aerobic organisms to generate energy
lac operon	series of genes required for transport and metabolism of lactose in some bacteria
Lagging strand	one stand of DNA which can only be replicated in short strands (Okazaki fragment) or discontinuously
Leading strand	one strand of DNA which can be replicated continuously
Lignin	complex organic polymer found in the cell wall of some plants and algae
Liposome	small sac of phospholipids enclosing a fluid droplet
Lithosphere	rigid outer part of the Earth consisting of upper mantle and crust
Macrophage	white cell of the immune system
Micelle	simple fat globule
Microbiome	the total amount of genes of all our microbes

Missense mutation	a point mutation changing a single nucleotide base resulting in a different amino acid being coded for
Mitochondria	energy generating organelle of the cell
Mole [not the animal]	unit for the amount of a substance that contains the same number of atoms or molecules as 12 grams of Carbon-12
Monolayer	single layer
mRNA	messenger RNA which codes for proteins
Mycelium	branching network produced by hyphae of fungi
NAD/NADH	a necessary 'helper' molecule found in all cells
Nitrogenases	an enzyme able to 'fix' atmospheric nitrogen
Nonsense mutation	a mutation coding in a 'stop' codon so producing a shorter, unfinished protein molecule
Nuclease	enzyme which degrades DNA
Nucleotides	one of the four bases which form DNA

Okazaki fragments	small DNA fragments created by a lagging strand
Organelle	small internal organ of a cell
Peptidoglycan	a polymer consisting of sugars and amino acids forming a cell wall in bacteria
Phosphodiester bond	a chemical bond joining successive sugars together to form a polynucleotide chain
Phospholipid	molecule with a phosphate head and fatty acid tail
Photoautotroph	organism gaining energy from sunlight
Photosynthesis	production of energy from sunlight
Pilus	hair like appendage on many bacteria
Plasmid	small circular double stranded DNA molecule found in bacterial cells
Prokaryote	single celled organism without organelles
Purine	organic molecule consisting of two rings making the base of two nucleotides Adenosine and Guanine

Pyrimidine	organic molecule consisting of a single ring structure making the base of two nucleotides Cytosine and Thymine
Ribosome	complex cellular machinery that builds proteins
RNA	polymer of nucleotides involved in the expression of genes into proteins
RNA primase	an enzyme that creates short lengths of RNA called primers
rRNA	ribosomal RNA - the RNA component of the ribosome
Saccharides	simple carbohydrates or sugars made from carbon, oxygen and hydrogen
Saprotroph	organism that feeds on decaying organic matter
Shannon-Weiner index	common calculation and measure of biodiversity
Svedberg unit	unit of time measuring sedimentation rate of particles based on size and shape

Thermotaxis	movement of an organism towards a heat source
Thymine	one of the four bases of DNA
Transconjugation	transfer of genetic material between bacteria
Transcription	first step of gene expression involving conversion of DNA into messenger RNA
Transduction	transfer of DNA from one bacteria to another
Translation	second step of gene expression involving translating messenger RNA into an amino acid chain
tRNA	transfer RNA carries an amino acid for translation of the mRNA by the ribosome into a protein
Ubiquinone	co-enzyme involved in many processes including the protein transport chain and ATP production
Uracil	nucleotide base used by RNA replacing thymine
Vacuole	space within the cytoplasm of a cell enclosed in a membrane

Index

16S rRNA, 85, 142
abiotic, 196, 227, 252
adenine, 27, 31, 71
adenosine triphosphate, ATP, 71
ADP, 71
aerobic respiration, 82
amino acids, 8, 19, 45, 89
ammonia, 22
AMP, 71
anaerobic respiration, 81
analogous, 111
ancient woodland, 269
Anthrax, 234
antibiotic, 168
antibiotic resistance, 171
antigen, 124, 300
ants, 240
archaea, 23, 81, 182
ash dieback, 211, 284
aspirin, 279
ATP synthase, 75
autotroph, 206
Avogadro's constant, 18
AZT, 135
bacteria, 23, 81, 84, 138, 154
bacteriophages, 302
beech tree, 278
Big Bang, 17
biodiversity, 153, 171, 195, 209, 210, 271, 303
biofuel, 295
biogeography, 227
biotic, 258
birch trees, 278
black smokers, 250
brine, 252
calculating biodiversity, 212
cap and tail, 44
CAP, catabolite activator protein, 59
carbon cycle, 228
codon, 48
community, 196
coppicing, 277
coral reefs, 261
CRISPR Cas9, 290
cyanide, 79
cyanobacteria, 81, 198, 235
cyclical AMP, cAMP, 59
Cystic fibrosis, 299
cytochrome b-c1, 79
cytochrome oxidase, 79
cytoplasm, 22
cytosine, 31
Darwin, 91
Darwin's finches, 99
dead wood, 279
degeneracy of the genetic code, 49
disulphide bridge, 20

DNA, 26
DNA helicase, 38, 39
DNA polymerase, 38
dominant species, 273
double helix, 36
double strand of DNA, 35
Duchenne muscular dystrophy, 69
E. coli, 56, 148, 192
ecological niche, 270
elongation phase, 53
endosymbiosis theory, 81
Enzyme-Linked Immunosorbent Assay or ELISA, 300
ether bonds, 183
eukaryote, 23, 24
evolution, 91, 119, 164
exons, 66
exponential growth, 204
fatty acid, 9
food web, 207, 259
fundamental niche, 270
fungi, 280
gene expression, 55
genes, 55
genes in HIV, 121
genetic drift, 114
genetic fingerprint, 296
genetically modified organisms, 288
geosim, 233
Gram staining, 139
guanine, 31
HAART, 137
habitat, 196
halophiles, 189
hedgerows, 275
Helicobacter pylori, 145
herd immunity, 178
Hill number, 220
histones, 33
HIV, 119
homologous, 112
hydrogen bond, 20, 34
hydrothermal vents, 250
interspecific competition, 273
intraspecific competition, 274
introns, 66
ionic bond, 20
kelp, 260
keystone species, 240, 258, 274
lac operon, 57
lacI, 59
lactose, 56
lagging strand, 40
leading strand, 40
lichen, 281
Linnaeus, 105
liposome, 12
logistic growth, 206
LUCA, 93, 113
macrophages, 122
major histocompatibility complex, 123
malaria, 64
Mendel, 102

methanogens, 189
micelle, 11
missense mutation, 62
mitochondria, 73
molecules essential to
life, 19
monoculture, 213
monosaccharide, 21
muscular dystrophy, 69
mutation, 62, 115, 130,
172
NAD, NADH, 76
natural selection, 98
niche, 196
Nitrification, 202
nitrogen, 265
nitrogen cycle, 198
Nobel Prize, 56, 146, 161,
169, 291
nuclease, 68
nucleotide, 27
nucleus, 36
oak tree, 277
obligate intracellular
bacteria, 88
Okazaki fragments, 42
oxygen levels, 93, 203,
256
pea plants, 102
penicillin, 168, 172
pentose ring, 27
phagocytosis, 123
phosphate group, 29
phospholipid, 8
phospholipid bilayer, 12
plantations, 270
point mutation, 65
poisoning, 79
polyA tail, 44
polycistronic, 68
polymer, 20
polymerase chain
reaction, 291
polysaccharide, 21
population, 196
prokaryote, 23
proteins, 20, 45
proton motive force, 80
resource partitioning,
274
respiration, 76
respiratory enzyme
complex, 77
rhododendron, 273
ribose, 19
ribose sugar, 28, 71
ribosome, 45, 50, 185
ribosome binding sites,
51
Rickettsia prowazekii, 88
RNA, 43
RNA primase, 38
salt water, 252
saprotroph, 207
sea temperature, 255
seed banks, 303
semi-conservative
replication, 38
setting the reading
frame, 52
Shannon-Weiner index,
217

sickle cell anaemia, 63
silent mutation, 64
Simpson index, 215
single stranded binding proteins, 39
SIV, 131
sliding clamp, 38
slime moulds, 238
sunlight zone, 253
surface area, 14
surface area to volume ratio, 15
TALEN, 290
termination phase, 53
tetanus, 176, 234
thymine, 31
tides, 247
transcription, 43
transfer RNA, tRNA, 45
translation, 50
triplet, 48
ubiquinone, 78
universal code, 48
uracil, 44
vaccines, 176
water pressure, 257
waxy ear canals, 62
weight of the earth, 18
white cell, 122
woodland management, 276
zinc finger nuclease, 290
β-galactosidase, 58
β-galactoside permease, 58

Lightning Source UK Ltd.
Milton Keynes UK
UKOW05f1824180916

283276UK00008B/120/P